Algebra for Symbolic Computation

Antonio Machì

Algebra for Symbolic Computation

 Springer

Antonio Machì
Department of Mathematics
University La Sapienza, Rome

Translated from the original Italian version by:
Daniele A. Gewurz
Department of Mathematics, University La Sapienza, Rome (Italy)

UNITEXT – La Matematica per il 3+2

ISSN print edition: 2038-5722   ISSN electronic edition: 2038-5757

ISBN 978-88-470-2396-3   ISBN 978-88-470-2397-0 (eBook)
DOI 10.1007/978-88-470-2397-0

Library of Congress Control Number: 2011945436

Springer Milan Heidelberg New York Dordrecht London

9 8 7 6 5 4 3 2 1

Cover-Design: Beatrice &, Milano

Typesetting with LaTeX: PTP-Berlin, Protago TeX-Production GmbH, Germany
(www.ptp-berlin.eu)

Springer-Verlag Italia S.r.l., Via Decembrio 28, I-20137 Milano
Springer is a part of Springer Science+Business Media (www.springer.com)

Preface

This book arose from a course of lectures given by the author in the universities of Paris VII and Rome "La Sapienza". It deals with classical topics in Algebra, some of which have been relegated for a long time to a marginal position, but have been brought to light by the development of the so-called symbolic computation, or computer algebra. I have tried to present them in such a way as to require only the very basic elements of Algebra, and often not even those, emphasising their algorithmic aspects. It is clear that a thorough comprehension of these subjects would be greatly simplified if it is accompanied by exercises at the computer. Many examples of algorithms that may be easily translated into computer programs are given in the text, others may be easily deduced from the theory. The literature on the subject is very rich; the bibliography at the end of the book only mentions the texts and the articles that I have consulted.

The first chapter of the book deals with elementary results like the Euclidean algorithm, the Chinese remainder theorem, and polynomial interpolation. The second chapter considers the p-adic expansion of rational and algebraic numbers and also of rational functions. The resultant of two polynomials is explained in the third chapter, where many applications are also given (for instance, augmented and reciprocal roots and Hurwitz polynomials). In the fourth chapter the problem of the polynomial factorisation is discussed; in particular, the Berlekamp method is studied in greater detail. Finally, in the fifth chapter we consider the Discrete and Fast Fourier Transform, and also its interpretation in terms of the representation theory of Abelian groups; the $n \log n$ complexity is also explained.

Every chapter is equipped with exercises, and some results are presented in this form. The text proper does not make use of them, except when specifically indicated.

I have the pleasure to thank the students that have sat in my classes over the years for their remarks and suggestions. I also want to thank Marina Avitabile for her careful reading of the book and her penetrating comments, and Daniele A. Gewurz for his clever remarks and excellent translating job. It goes without saying that I remain the only responsible for possible errors or lack of clarity.

Rome, September 2011   *Antonio Machì*

Contents

1

The Euclidean algorithm, the Chinese remainder theorem and interpolation

1.1 The Euclidean algorithm

Let m be an arbitrary integer number (positive, negative or zero), n a positive integer, and let

$$\ldots, -kn, \ldots, -2n, -n, 0, n, 2n, \ldots, kn, \ldots$$

the set of multiples of n. There exist two consecutive terms of this sequence, qn and $(q+1)n$, such that:

$$qn \leq m < (q+1)n. \tag{1.1}$$

Let r be the difference

$$r = m - qn.$$

The operation consisting in finding the two numbers q and r is called *division* of m (dividend) by n (divisor). The integer q is the *quotient* (the greatest integer whose product by n is not greater than m, and as such it is uniquely determined), and r is the *remainder* of the division (which is uniquely determined as well). From (1.1), by subtracting qn from each term, we get

$$0 \leq r < n$$

and hence: *the remainder is never negative and is less than the divisor*. If $r = 0$, the number m is said to be a *multiple* of n, and n is said to be a *divisor* of m, and we write $n|m$. If $m \neq 1$ has no divisors but itself and 1, then m is a *prime* number.

Consider now two integers m and n, $m \geq n > 0$, and look for the integers that divide both m and n. We shall see that the problem reduces to finding the divisors of *a single integer d*. Divide m by n:

$$m = qn + r \qquad\qquad 0 \leq r < n.$$

Machì A.: Algebra for Symbolic Computation.
DOI 10.1007/978-88-470-2397-0_1, © Springer-Verlag Italia 2012

It follows that $r = m - qn$. It is then apparent that an integer dividing both m and n divides r too. On the other hand, if an integer divides n and r, the same equality tells us that that integer divides m too, so it divides both m and n. In other words, the common divisors of m and n coincide with those of n and r. Now, dividing n by r, we deduce as above that the common divisors of n ad r (hence those of m and n) are those of r and of the remainder of the division. Going on this way, we get a sequence of divisions that eventually reach a remainder equal to zero, since the remainders are strictly decreasing non-negative integers (we have set $q = q_1$ and $r = r_1$):

$$m = nq_1 + r_1, \qquad r_1 < n,$$
$$n = r_1 q_2 + r_2, \qquad r_2 < r_1,$$
$$r_1 = r_2 q_3 + r_3, \qquad r_3 < r_2,$$
$$\vdots$$
$$r_{k-3} = r_{k-2} q_{k-1} + r_{k-1}, \qquad r_{k-1} < r_{k-2},$$
$$r_{k-2} = r_{k-1} q_k,$$

with the kth remainder r_k equal to zero. By the above argument we have now that the common divisors of m and n are the common divisors of r_{k-1} (the last non-zero remainder) and 0 and, since all integers divide 0, we have that *the common divisors of m and n are exactly the divisors of r_{k-1}*. Having set $d = r_{k-1}$, we write $d = (m, n)$ or $d = \gcd(m, n)$. Now, d divides both m and n, and since among the divisors of d there is d itself, d is the greatest among the mutual divisors of m and n, hence the name of *greatest common divisor* of m and n. This name, however, conceals the main property of this number, lying not only in being the greatest among the divisors of both m and n, but in having as its divisors all the divisors of m and n.

So we have the following *Euclidean algorithm*:

input: m, n;
output: $\gcd(m, n)$;

$u := m$;
$v := n$;
while $v \neq 0$ do:
$q := \mathrm{quotient}(u, v)$;
$t := u$;
$u := v$;
$v := t - qv$.

If $d = (m, n) = 1$, then m and n have no common divisors different from 1: they are *coprime* or *relatively prime*.

The remainder $r_1 = m - qn$ is a linear combination of m and n:

$$r_1 = 1 \cdot m + (-q_1) \cdot n,$$

and this holds for r_2 too: we have $r_2 = n - q_2 r_1 = n - q_2 m + q_1 q_2 n$; hence:

$$r_2 = (-q_2) \cdot m + (1 + q_1 q_2) \cdot n.$$

This happens for all the remainders r_i, $i = 1, 2, \ldots, k$, so it also holds for $r_{k-1} = d$. Hence, we have:

Theorem 1.1 (The Bézout identity). *Given two integers m and n, there exist two integers h and k such that $d = (m, n)$ is a linear combination of m and n: $d = hm + kn$.*

Remark. The integers h and k in Theorem 1.1 are not uniquely determined. Indeed, for every integer s we have $d = (h \pm sn)m + (k \mp sm)n$.

Let $m, n > 1$; note that in the Bézout identity it is not possible that h and k are both positive, nor both negative. Assume now $k > 0$; then it is always possible to find h' and k' with $|h'| < n$ and $k' < m$ such that $h'm + k'n = d$. Indeed, if $k < m$, $hm = d - kn$, and hence $|h|m = kn - d$, and if we had $|h| \geq n$, then $nm \leq |h|m = kn - d < mn - d$, a contradiction. If $k \geq m$, divide k by m: $k = mq + r$, $0 \leq r < m$. Then $(h + qn)m + (k - mq)n = d$, and setting $h' = h + qn, k' = k - qm$ from $k' < m$ it follows, by the above, that $h' < n$.

By modifying the program for the Euclidean algorithm we get the *Bézout algorithm*. It provides the triple $[h, k, d]$:

input: m, n;
output: h, k, d;
$u := [1, 0, m]$;
$v := [0, 1, n]$;
while $v_3 \neq 0$ do:
$q := \text{quotient}(u_3, v_3)$;
$t := u$;
$u := v$;
$v := t - qv$

(u_3 and v_3 denote the third component of u and v).

The input of the algorithm consists of the two integers m and n. In the form given by Bézout's theorem they are written $m = 1 \cdot m + 0 \cdot n$ and $n = 0 \cdot m + 1 \cdot n$, which yield the starting triples for the algorithm ($m = (m, m)$ and $n = (n, n)$).

An integer m divided by n gives a non-negative remainder that is less than n. So the possible remainders are the integers $0, 1, \ldots, n - 1$ (and all of them can be obtained, since a number m less than n, divided by n, gives m as remainder, the quotient being 0). If the remainder is r, we say that m *is congruent to r modulo n*, and we write $m \equiv r \bmod n$. As a last thing, let us recall the *fundamental theorem of arithmetic*: *every integer number different from 1 is either prime or can be written as a product of prime factors and this,* *up to reordering the factors, can be done in a unique way.* (The uniqueness of this decomposition is another consequence of the existence of the greatest common divisor; see Exercise 3, below.)

Exercises

1. Prove that if four numbers m, n, q, r are such that $m = qn + r$ and $0 \leq r < n$, then q and r are the quotient and the remainder of the division of m by n.

2. Let $m|ab$ and $(m, a) = 1$. Prove that in this case $m|b$.

3. Prove that $(m, ab) = 1$ if and only if $(m, a) = 1$ and $(m, b) = 1$. Extend this result to the case of several integers. Use this fact to prove that the factorisation of an integer as a product of primes is unique.

4. Let $(m, n) = 1$. Prove that if $hm + kn = 1$ and $h'm + k'n = 1$, then $h' \equiv h \bmod n$ and $k' \equiv k \bmod m$.

5. Let $a_1, a_2, \ldots, a_n$ be positive integers, and let a_n be the smallest one. Let r_i be the remainder of the division of a_i by a_n, $i \neq n$. Prove that:

$$\gcd(a_1, a_2, \ldots, a_n) = \gcd(r_1, r_2, \ldots, r_{n-1}, a_n).$$

(Note that the smallest number of the first n-tuple is the greatest of the second one.) By repeating this operation for the second n-tuple, and so on, we get to a n-tuple in which all elements are equal to 0, except one. Prove that this element is the $\gcd(a_1, a_2, \ldots, a_n)$.

6. In the *Fibonacci sequence* 0, 1, 1, 2, 3, 5, 8, 13, 21,... each term, starting from the third one, is obtained by summing the two terms that precede it:

$$f_{n+2} = f_{n+1} + f_n, \quad n \geq 0, \quad f_0 = 0, \quad f_1 = 1.$$

i) Prove that any two consecutive Fibonacci numbers f_{k+1} and f_k are relatively prime.

ii) Prove that the Euclidean algorithm, applied to f_{n+2} and f_{n+1} stops after exactly n steps. [*Hint*: note that, if we write the defining relation for Fibonacci numbers as $f_{n+2} = f_{n+1} \cdot 1 + f_n$, we have, since $f_n \leq f_{n+1}$, that f_n is the remainder of the division of f_{n+2} by f_{n+1} (and the quotient is 1).]

iii) (Lamé Theorem) Prove that if $u > v > 0$ are such that the Euclidean algorithm for u and v ends after exactly n steps, and u is the least number with this property, then $u = f_{n+2}$ and $v = f_{n+1}$. [*Hint*: note that $r_{n-2} \geq 2 = f_3$, so $r_{n-3} \geq 2 + 1 = 3 = f_4$, and that in general every remainder is greater or equal than the sum of the two remainders following it.]

7. Given two integers m and n, if we substitute in the definition of gcd the word "least" for "greatest" and "multiple" for "divisor", we get the "dual" definition of a *least common multiple*: it is *the number μ such that the common multiples of m and n are exactly the multiples of μ.* We write $\mu = \text{lcm}(m, n)$.

i) Prove that μ exists and is equal to $mn/(m, n)$.

ii) Extend the above to the case of more than two integers.

8. Prove that, for every n, none of the numbers

$$n! + 2, n! + 3, \ldots, n! + n$$

is prime. Do the same taking $\operatorname{lcm}(1, 2, \ldots, n)$ in the place of $n!$. (So, if we consider the largest prime number less than $n! + 2$ and the smallest prime greater than $n! + n$, we find that in the sequence of prime numbers there are arbitrarily long gaps.)

1.2 The Chinese remainder theorem and Lagrange numbers

Take now two pairs of numbers m, n and a, b and consider the following question: is there an integer c that divided by m gives a as remainder, and divided by n gives b as remainder? Such an integer does not necessarily exist. For instance, taking $m = 4$, $n = 6$, $a = 1$ and $b = 2$ we would have $c = 4q_1 + 1 = 6q_2 + 2$, that is, c odd and even simultaneously. With $a = 2$ and $b = 4$ we have the solution $c = 10$. Indeed, $10 = 4 \cdot 2 + 2 = 1 \cdot 6 + 4$.

Thus, the existence of a solution depends on the choice of the parameters. Given m and n, a sufficient condition for a solution to exist independently of a and b is that m and n be relatively prime, as the following theorem shows.

Theorem 1.2 (Chinese remainder theorem). *Let m and n be relatively prime integers. Then, for all pairs a, b of integers, there exists an integer c such that*

$$c \equiv a \bmod m \quad and \quad c \equiv b \bmod n.$$

Moreover, this solution is unique $\bmod$ *mn (that is, c is the only integer greater or equal than 0 and less than mn that solves both congruences).*

Proof. Since $(m, n) = 1$, there exist h and k such that $hm + kn = 1$. Hence we have $kn \equiv 1 \bmod m$ and $hm \equiv 1 \bmod n$. If $a = 1$ and $b = 0$ we already have the solution: $c = kn$ (and, for $a = 0$ and $b = 1$, the solution $c = hm$). In general, we claim that $c = bhm + akn$ is the required integer. Indeed, modulo m, we have $c \equiv akn \equiv a \cdot 1 = a \bmod m$, and, modulo n, $c \equiv bhm \equiv b \cdot 1 = b \bmod n$. As for uniqueness, if d is another solution, since d is congruent to a and b modulo m and n, respectively, we have $d \equiv c$ modulo m and n, so m and n divide $d - c$. But, as m and n are relatively prime, their product mn divides $d - c$ too, that is, $d \equiv c \bmod mn$. So the solution modulo mn is unique. In other words, the integer c with $0 \le c < mn$ that solves the two congruences exists and is unique. All other solutions can be obtained by adding to c a multiple of mn. $\diamond$

Remark. As a consequence, the integers a and b in the theorem turn out to be the coefficients needed to write c as a linear combination of hm and kn. Note further that the example above, with $m = 4$, $n = 6$, $a = 2$ and $b = 4$, in addition to the solution $c = 10$ also admits the solution $c = 22 = 4 \cdot 5 + 2 = 6 \cdot 3 + 4$, and 10 and 22 are not congruent modulo $4 \cdot 6 = 24$. Hence, if m and n are not relatively prime, not only a solution may not exist, but if it exists it may well not be unique.

Example. 1. Let us prove that the group $Z_3 \oplus Z_4$ is isomorphic to the group Z_{12}. The former consists of the pairs (x, y) with $x = 0, 1, 2$ and $y = 0, 1, 2, 3$, with the operation

$$(x_1, y_1) + (x_2, y_2) = ((x_1 + x_2) \bmod 3, (y_1 + y_2) \bmod 4);$$

the latter consists of the integers $0, 1, \ldots, 11$ with the sum mod 12. Given a pair $(x, y) \in Z_3 \oplus Z_4$, by the Chinese remainder theorem the integer c, $0 \leq c < 3 \cdot 4 = 12$, congruent to $x \bmod 3$ and to $y \bmod 4$, exists and is unique. The correspondence $(x, y) \to c$ is the isomorphism required. More generally, for $(m, n) = 1$, we have that $Z_m \oplus Z_n$ is isomorphic to Z_{mn}.

2. Let $\varphi(n)$ be *Euler's (totient) function* defined as follows:

i) $\varphi(1) = 1$;

ii) $\varphi(n) =$ number of the integers less than n and coprime with n.

Let us show that if $(m, n) = 1$ then $\varphi(mn) = \varphi(m)\varphi(n)$. Denote by C_n the set (actually, the group) of the integers less than n and coprime with n. If $(x, m) = (y, n) = 1$, then the integer c of the Chinese remainder theorem, that is, the integer congruent to x and to y modulo m e n, respectively, is coprime with m and n, and so with the product mn too. Then the correspondence $C_m \times C_n \longrightarrow C_{mn}$ given by $(x, y) \to c$ is well defined by the uniqueness of c. Conversely, if $c \in C_{mn}$, then by taking as x and y the remainders of the division of c by m and by n we find that the inverse correspondence $c \to (x, y)$ is well defined (uniqueness of the remainder). So we have a bijection.

3. If $(m, n) = 1$, we have $am + bn = 1$, for suitable a and b. Set $e = am$, $e' = bn$. We have:

$$e^2 = (am)^2 = am \cdot am = am(1 - bn) = am - abmn \equiv am = e \bmod mn,$$

and analogously $e'^2 \equiv e' \bmod mn$. This fact is expressed by saying that, in the ring of congruence classes modulo mn, e and e' are *idempotent*. Moreover, they are *orthogonal*:

$$ee' = am \cdot bn = abmn \equiv 0 \bmod mn,$$

and their sum is 1: $e + e' = 1$.

The Chinese remainder theorem can be generalised to several moduli.

Theorem 1.3. *Let $m_0, m_1, \ldots, m_n$ be pairwise coprime integers, and let u_0, $u_1, \ldots, u_n$ be arbitrary integers. Then there is an integer u such that*

$$u \equiv u_i \bmod m_i,$$

$i = 0, 1, \ldots, n$, *and this u is unique modulo the product $m_0 m_1 \cdots m_n$. In other words, solving the system of congruences $x \equiv u_i \bmod m_i$, $i = 0, 1, \ldots, n$, is equivalent to solving the single congruence $x \equiv u \bmod m_0 m_1 \cdots m_n$.*

We shall give two proofs for this theorem, one just establishing the existence and uniqueness, and a constructive one.

First proof. We prove first that if u exists, then it is unique modulo the product $m = m_0 m_1 \cdots m_n$. Indeed, if $v \equiv u_i \bmod m_i$, for all i, then $u - v \equiv 0 \bmod m_i$, so (the m_i are pairwise coprime) $u - v \equiv 0 \bmod m$, or $u \equiv v \bmod m$. Consider now, for all $u = 0, 1, \ldots, m - 1$, the $(n + 1)$-tuple $(u \bmod m_0, u \bmod m_1, \ldots, u \bmod m_n)$. Those $(n + 1)$-tuples are, for different us, all different, since if for some u and v in $\{0, \ldots, m - 1\}$ we had:

$$(u \bmod m_0, \ldots, u \bmod m_n) = (v \bmod m_0, \ldots, v \bmod m_n),$$

then $u \equiv v \bmod m$. But the $(n+1)$-tuples $(v_0, v_1, \ldots, v_n)$, where $0 \leq v_i < m_i$, are also all different, and their number is equal to that of the former tuples, that is, m. So for a given $(n+1)$-tuple $(u_0, u_1, \ldots, u_n)$ there is one of the form $(u \bmod m_0, u \bmod m_1, \ldots, u \bmod m_n)$ equal to it.

Second proof. This proof extends the one for two integers we have seen above, and just like in that case we give a method to construct the solution. Set $m = m_0$ and $n = m_1 m_2 \cdots m_n$: so there exist two integers h and k, with $k < m$ (cf. the remark after Theorem 1.1), such that $hm + kn = 1$; hence, having set $L_0 = kn$,

$$L_0 \equiv 1 \bmod m,$$
$$L_0 \equiv 0 \bmod n.$$

L_0 is divisible by the product $m_1 m_2 \cdots m_n$, so in particular by m_i, $i = 1, 2, \ldots, n$. From this follows:

$$L_0 \equiv 1 \bmod m_0,$$
$$L_0 \equiv 0 \bmod m_i,$$

$i = 1, 2, \ldots, n$. (Note that L_0 solves the problem for the $(n + 1)$-tuple $u_0 = 1, u_1 = u_2 = \ldots = u_n = 0$.) Applying the same argument to the remaining m_is we find $n + 1$ integers $L_0, L_1, \ldots, L_n$ such that:

$$L_k \equiv 1 \bmod m_k,$$
$$L_k \equiv 0 \bmod m_j,$$

for $j \neq k$. The integer $u_0 L_0 + u_1 L_1 + \ldots + u_n L_n$ is a common solution to the congruences given. For uniqueness, see the first proof. $\diamond$

Remark. The L_ks form a "mixed basis" for the integers mod m, where $m = m_0 m_1 \cdots m_n$, in the sense that for each such integer u there is a unique $(n+1)$-tuple $(u_0, u_1, \ldots, u_n)$, with $0 \leq u_k < m_k$ such that $u = \sum_{k=0}^{n} u_k L_k$. This $(n+1)$-tuple can be considered a representation of u: we shall call it the *modular representation* of u, with respect to the moduli $m_0, m_1, \ldots, m_n$. By representing an integer in this way no information is lost since, by the Chinese remainder theorem, we may uniquely find back u from the u_is.

We shall call the L_ks *Lagrange numbers*.

Let A_1, A_2, ..., A_r be unitary commutative rings. The *direct sum*

$$A = A_1 \oplus A_2 \oplus \cdots \oplus A_r,$$

is the ring having as elements the r-tuples $(a_1, a_2, \ldots, a_r)$, and as operations the sum and product of these r-tuples defined componentwise. The r-tuples $(0, 0, \ldots, 0, a_k, 0, \ldots, 0)$ form a subring $\overline{A}_k$ isomorphic to A_k. The unit element of A is the r-tuple $1 = (1, 1, \ldots, 1)$, and the one of $\overline{A}_k$ is $e_k = (0, 0, \ldots, 0, 1, 0, \ldots, 0)$. So the sum of the e_ks is 1, the unit of A:

$$e_1 + e_2 + \cdots + e_r = 1,$$

while the product of e_i and e_j, with $i \neq j$ is zero:

$$e_i e_j = 0;$$

moreover:

$$e_k^2 = e_k.$$

Hence, the units of the A_ks are orthogonal idempotents. Given an element $a = (a_1, a_2, \ldots, a_k, \ldots, a_r) \in A$, it is the sum of its components $\overline{a}_k = (0, 0, \ldots, 0, a_k, 0, \ldots, 0)$; but $(0, 0, \ldots, 0, a_k, 0, \ldots, 0) = ae_k$, so:

$$\overline{a}_k = ae_k.$$

Hence, $A = \overline{A}_1 \oplus \overline{A}_2 \oplus \cdots + \oplus \overline{A}_r$, that is, A is the direct sum of its subrings $\overline{A}_k$ (*internal* direct sum).

Conversely, let A be a unitary commutative ring such that the unit 1 admits a partition into orthogonal idempotents: $e_1 + e_2 + \cdots + e_r = 1$. Then the elements ae_k, $a \in A$, form a subring since:

$$ae_k \pm be_k = (a \pm b)e_k,$$
$$(ae_k)(be_k) = abe_k^2 = abe_k.$$

Denote by $\overline{A}_k$ this subring, and by $\overline{a}_k$ the element ae_k. $\overline{A}_k$ has e_k as its unit:

$$\overline{a}_k e_k = ae_k e_k = ae_k^2 = ae_k = \overline{a}_k,$$

and the elements $a \in A$ admit the decomposition

$$a = a \cdot 1 = a(e_1 + e_2 + \cdots + e_r) = ae_1 + ae_2 + \cdots + ae_r = \overline{a}_1 + \overline{a}_2 + \cdots + \overline{a}_r,$$

with $ae_k \in \overline{A}_k$, where we have set $ae_k = \overline{a}_k$. This decomposition is unique; indeed, if we had

$$a = x_1 e_1 + x_2 e_2 + \cdots + x_r e_r,$$

with $x_k \in A$ and $x_k e_k \in \overline{A}_k$, then, multiplying by e_j we would find $x_j e_j = ae_j$. Hence, $A = \overline{A}_1 \oplus \overline{A}_2 \oplus \cdots + \oplus \overline{A}_r$.

In Example 3 above we saw that $e = L_0$ and $e' = L_1$ are orthogonal idempotents and sum to 1. This holds for all L_ks. Having set $m = m_0 m_1 \cdots m_n$, we have:

1. $L_0 + L_1 + \ldots + L_n \equiv 1 \bmod m$.

   Taking $u_0 = u_1 = \ldots = u_n = 1$ we have $\sum L_i \equiv 1 \bmod m_i$. It follows that $\sum L_i - 1$ is divisible by m_i for all i, so it is divisible by their product m.

2. $L_i L_j \equiv 0 \bmod m, i \neq j$.

   This can be seen by observing that the product $L_i L_j$ includes as its factors both $\prod_{s \neq i} m_s$ and $\prod_{s \neq j} m_s$, hence all m_is and their product m.

3. $L_k^2 \equiv L_k \bmod m$.

   It follows from 1. that

$$L_k = L_k \cdot 1 \equiv L_k \cdot (L_0 + L_1 + \ldots + L_k + \ldots + L_n) \bmod m.$$

By 2., all the products $L_k L_j, j \neq k$, are zero modulo m and L_k^2 is the only remaining summand.

Let now $A = Z_m$ be the ring of congruence classes modulo m, with $m = m_0 \cdots m_n$. By the above, the L_ks make it possible to decompose A as a direct sum of subrings $A_k = \{a L_k \bmod m, a \in A\}$. Having set $\bar{a}_k = a L_k \bmod m$ (so $a L_k = mq + \bar{a}_k$, for some q), the element a (the congruence class of $a \bmod m$) can be decomposed as

$$a \equiv \bar{a}_1 + \bar{a}_2 + \cdots + \bar{a}_r \bmod m,$$

and since $L_k \equiv 1 \bmod m_k$ and $L_k \equiv 0 \bmod m_i$, $i \neq k$, we have:

$$\bar{a}_k \equiv a \bmod m_k, \bar{a}_k \equiv 0 \bmod m_i, i \neq k.$$

The subring A_k is isomorphic to the ring of congruence classes modulo m_k. Indeed, consider the correspondence $Z \longrightarrow A_k$ given by $a \to \bar{a}_k$. Clearly, this is a surjective homomorphism. If $\bar{a}_k = 0$, then $a L_k$ is divisible by m, so by all the m_is, and in particular by m_k; but, since $L_k \equiv 1 \bmod m_k$, m_k has to divide a, that is $a \equiv 0 \bmod m_k$. Conversely, if $a \equiv 0 \bmod m_k$ then $a L_k \equiv 0 \bmod m_k$; but, as $a L_k \equiv 0 \bmod m_i$, for $i \neq k$ (since this holds for the L_ks), we know that $a L_k$ is divisible by all the m_is, and hence by m. In other words, $\bar{a}_k \equiv 0 \bmod m$. So, in this homomorphism the elements of the kernel are exactly the multiples of m_k, and by standard results about homomorphisms we have $A_k \simeq Z/m_k Z = Z_{m_k}$.

In conclusion,

$$Z_m = A_0 \oplus A_1 \oplus \cdots \oplus A_n$$

(internal direct sum), where the component in A_k of $a \in Z_m$ is the remainder of the division of a by m_k.

Example. For $A = Z_{12}$, we have $1 \equiv 9 + 4 \bmod 12$ (this relation corresponds to the identity $1 \equiv -1 \cdot 3 + 1 \cdot 4$ yielded by Bézout in Z). We have $9 \cdot 4 = 36 \equiv 0 \bmod 12$, $9^2 = 81 \equiv 9 \bmod 12$, $4^2 = 16 \equiv 4 \bmod 12$, so with $e_1 = 9$ and $e_2 = 4$ we have two orthogonal idempotents summing to 1. Multiplying all the elements of Z_{12} by 9 we get $A_1 = \{0, 9, 6, 3\}$. (If $n = 4k + h$, then $n \cdot 9 = 4k \cdot 9 + h \cdot 9 \equiv h \cdot 9 \bmod 12$, so multiplying by 9 yields four distinct elements: $0 \cdot 9$, $1 \cdot 9$, $2 \cdot 9 \equiv 6 \bmod 12$, $3 \cdot 9 \equiv 3 \bmod 12$.) Analogously, $A_2 = \{0, 4, 8\}$, so

$$Z_{12} = \{0, 9, 6, 3\} \oplus \{0, 4, 8\}.$$

Note that the product of an element in A_1 by one in A_2 is equal to zero. Every element of Z_{12} can be written uniquely as a sum $a_1 + a_2$, $a_1 \in A_1$, $a_2 \in A_2$. For instance, $7 = 3 + 4$, $5 = 9 + 8$, and the sum and the product are computed componentwise:

$$7 + 5 = (3 + 9) + (4 + 8) = 12 + 12 = 24 \equiv 0 \bmod 12,$$

and

$$7 \cdot 5 = (3 + 4) \cdot (9 + 8) = 3 \cdot 9 + 4 \cdot 8 = 59 \equiv 11 \bmod 12.$$

Now, $\{0, 9, 6, 3\}$ (the additive group generated by 9) is isomorphic to the ring $Z_4 = \{0, 1, 2, 3\}$ of congruence classes modulo 4, and $\{0, 4, 8\}$ (the additive group generated by 4) to the ring $Z_3 = \{0, 1, 2\}$, so we also have

$$Z_{12} \simeq \{0, 1, 2, 3\} \oplus \{0, 1, 2\}.$$

Consider now the product $7 \cdot 5$ under this decomposition of Z_{12}. The isomorphism gives

$$7 \to (7 \bmod 4, 7 \bmod 3) = (3, 1) \quad \text{and} \quad 5 \to (5 \bmod 4, 5 \bmod 3) = (1, 2),$$

so the product $7 \cdot 5 = (3, 1) \cdot (1, 2) = (3 \cdot 1, 1 \cdot 2) = (3, 2)$. Which number is congruent to 3 mod 4 and to 2 mod 3? The Chinese remainder theorem immediately yields the answer -1, that is, 11 mod 12, and we get again the previous result.

If $1 = e_1 + e_2 \cdots + e_n$ in Z_m, then A_k is the set of elements obtained by multiplying e_k by $0, 1, \ldots, \frac{m}{(m, e_k)} - 1$ (which are the remainders of the division by $\frac{m}{(m, e_k)}$; in the example above, with $e_1 = 9$, we get $\frac{12}{(12, 9)} = \frac{12}{3} = 4$, so we obtain all the elements of A_1 multiplying e_1 by 0, 1, 2, and 3). It is straightforward to see that $Z / \frac{m}{(m, e_k)} \simeq A_k$ in the isomorphism that maps the congruence class $h \bmod \frac{m}{(m, e_k)}$ to $h e_k$ (if $h_1 e_k = h_2 e_k$ then $(h_1 - h_2) e_k = 0$, so $(h_1 - h_2) e_k = t \frac{m}{(m, e_k)}$; hence, $\frac{m}{(m, e_k)}$ divides $h_1 - h_2$, being relatively prime with e_k, so $h_1 \equiv h_2 \bmod \frac{m}{(m, e_k)}$).

Exercise

9. Consider the analogue of the previous example for $A = Z_{30}$, with respect to the factorisations $30 = 2 \cdot 3 \cdot 5$ and $30 = 6 \cdot 5$.

Let us come back to the Chinese remainder theorem. To compute u we may proceed in any of two ways:

1. As seen in theorem: given the m_ks, compute the L_ks once for all and then, for each choice of the u_is, determine u as a linear combination of the L_ks with the u_is as coefficients. This is *Lagrange method*.

2. Consider together the m_is and the u_is as given, and recursively construct u starting from m_0 and u_0, then m_0, m_1 and u_0, u_1, and so on. This is *Newton method*.

Let us see now these two methods in detail.

1.2.1 Computing L_k and u (Lagrange's method)

Let $m_0, m_1, \ldots, m_n$ be given, pairwise relatively prime, natural numbers. So, for every pair (m_i, m_k), there exist $s_i^{(k)}$ and $s_k^{(i)}$ such that:

$$s_i^{(k)} m_i + s_k^{(i)} m_k = 1,$$

or:

$$s_i^{(k)} m_i = 1 - s_k^{(i)} m_k.$$

Hence all quantities $s_i^{(k)} m_i$, $i \neq k$, are congruent to 1 modulo m_k, as is their product L_k':

$$L_k' = \prod_{i \neq k} s_i^{(k)} m_i \equiv 1 \bmod m_k.$$

For $i \neq k$ this product has m_i as one of its factor, so:

$$L_k' \equiv 0 \bmod m_i, \ i \neq k.$$

In order to work with smaller numbers, let us transform L_k' as follows:

$$L_k' = \prod_{i \neq k} s_i^{(k)} m_i = \prod_{i \neq k} s_i^{(k)} \cdot \prod_{i \neq k} m_i,$$

and substitute $\prod_{i \neq k} s_i^{(k)}$ with the remainder it gives when divided by m_k. This leaves the congruence unaffected since, if $\prod_{i \neq k} s_i^{(k)} = qm_k + r$ then:

$$L_k' = (qm_k + r) \prod_{i \neq k} m_i = qm_k \prod_{i \neq k} m_i + r \prod_{i \neq k} m_i$$

$$\equiv r \prod_{i \neq k} m_i \bmod m_k.$$

Set $L_k = r \prod_{i \neq k} m_i$. As $L_k \equiv L'_k \equiv 1 \bmod m_k$, we have:

$$L_k \equiv 1 \bmod m_k,$$
$$L_k \equiv 0 \bmod m_i, \quad i \neq k.$$

To sum up, in order to compute L_k:

1. For each i, determine two integers $s_i^{(k)}$ e $s_k^{(i)}$ such that:

$$s_i^{(k)} m_i + s_k^{(i)} m_k = 1.$$

2. Multiply all $s_i^{(k)}$s, $i \neq k$, divide this product by m_k and take the remainder r_k.

3. Hence $L_k = r_k \prod_{i \neq k} m_i$.

These L_ks coincide with the numbers $b \prod_{i \neq k} m_i$ found in the proof of Theorem 1.3. Indeed, $b \prod_{i \neq k} m_i = r_k \prod_{i \neq k} m_i \bmod m_k$, as both are congruent to 1 mod m_k. Since $(\prod_{i \neq k} m_i, m_k) = 1$, we have $b \equiv r_k \bmod r_k$ and consequently $b = r_k$ as both are less than m_k.

Example. Let $m_0 = 7$, $m_1 = 11$, $m_2 = 13$, $m_3 = 15$. Proceeding as described above, we have:

$$\underset{-3}{s_0^{(1)}} \cdot 7 \ + \underset{2}{s_1^{(0)}} \cdot 11 = 1,$$

$$\underset{2}{s_0^{(2)}} \cdot 7 \ + \underset{-1}{s_2^{(0)}} \cdot 13 = 1,$$

$$\underset{-2}{s_0^{(3)}} \cdot 7 \ + \underset{1}{s_3^{(0)}} \cdot 15 = 1,$$

$$\underset{6}{s_1^{(2)}} \cdot 11 + \underset{-5}{s_2^{(1)}} \cdot 13 = 1,$$

$$\underset{-4}{s_1^{(3)}} \cdot 11 + \underset{3}{s_3^{(1)}} \cdot 15 = 1,$$

$$\underset{7}{s_2^{(3)}} \cdot 13 + \underset{-6}{s_3^{(2)}} \cdot 15 = 1.$$

With these values, we have:

$$L_0 = s_1^{(0)} s_2^{(0)} s_3^{(0)} \ (\mathrm{mod}\ 7) \cdot 11 \cdot 13 \cdot 15$$
$$= 2 \cdot -1 \cdot 1 \ (\mathrm{mod}\ 7) \cdot 11 \cdot 13 \cdot 15$$
$$= -2 \ (\mathrm{mod}\ 7) \cdot 2145$$
$$= -4290.$$

$$L_1 = s_0^{(1)} s_2^{(1)} s_3^{(1)} \ (\mathrm{mod}\,11) \cdot 7 \cdot 13 \cdot 15$$
$$= -3 \cdot -5 \cdot 3 \ (\mathrm{mod}\,11) \cdot 1365$$
$$= 45 \ (\mathrm{mod}\,11) \cdot 1365$$
$$= 1 \cdot 1365$$
$$= 1365.$$

$$L_2 = s_0^{(2)} s_1^{(2)} s_3^{(2)} \ (\mathrm{mod}\,13) \cdot 7 \cdot 11 \cdot 15$$
$$= 2 \cdot 6 \cdot -6 \ (\mathrm{mod}\,13) \cdot 1155$$
$$= -72 \ (\mathrm{mod}\,13) \cdot 1155$$
$$= -7 \cdot 1155$$
$$= -8085.$$

$$L_3 = s_0^{(3)} s_1^{(3)} s_2^{(3)} \ (\mathrm{mod}\,15) \cdot 7 \cdot 11 \cdot 13$$
$$= -2 \cdot -4 \cdot 7 \ (\mathrm{mod}\,15) \cdot 1001$$
$$= 56 \ (\mathrm{mod}\,15) \cdot 1001$$
$$= -4 \cdot 1001 = -4004.$$

Choose now four integers $u_0 = -1$, $u_1 = -2$, $u_2 = 2$, $u_3 = 6$; an integer congruent to u_i mod m_i, $i = 0, 1, 2, 3$, is

$$4290 - 2 \cdot 1365 - 2 \cdot 8085 - 6 \cdot 4004 = -38634,$$

and the unique solution modulo $m_0 m_1 m_2 m_3 = 7 \cdot 11 \cdot 13 \cdot 15 = 15015$ can be obtained by summing -38634 and $3 \cdot 15015$. We find $u = 6411$.

As seen above, the four integers L_k just found yield a decomposition of the ring Z_m, with $m = 15015$, as a direct sum of four subrings. If $a \in Z_m$, its components in the four summands are $aL_k \ \mathrm{mod}\ m$. For instance, let $a = 156$; then:

0. $\bar{a}_0 = aL_0 \ \mathrm{mod}\ 15015 = 156 \cdot -4290 \ \mathrm{mod}\ 15015 = 6435;$
1. $\bar{a}_1 = aL_1 \ \mathrm{mod}\ 15015 = 156 \cdot 1365 \ \mathrm{mod}\ 15015 = 2730;$
2. $\bar{a}_2 = aL_2 \ \mathrm{mod}\ 15015 = 156 \cdot -8085 \ \mathrm{mod}\ 15015 = 0;$
3. $\bar{a}_3 = aL_3 \ \mathrm{mod}\ 15015 = 156 \cdot -4004 \ \mathrm{mod}\ 15015 = 6006.$

The four numbers we have found sum up to 15171 which is exactly 156 mod 15015. Moreover, $\bar{a}_k \equiv a \ \mathrm{mod}\ m_i$; indeed: $6435 - 156 = 897 \cdot 7$; $2730 - 156 = 234 \cdot 11$; $0 - 156 = -12 \cdot 13$; $6006 - 156 = 390 \cdot 15$.

1.2.2 Computing u (Newton's method)

In Newton's method, that we are now about to discuss, the solution u is recursively constructed by using both the m_is and the u_is. If $u^{(k-1)}$ is a solution for $m_0, m_1, \ldots, m_{k-1}$ and $u_0, u_1, \ldots, u_{k-1}$, the method gives a solution $u^{(k)}$ for $m_0, m_1, \ldots, m_{k-1}, m_k$ and $u_0, u_1, \ldots, u_{k-1}, u_k$. Let $k = 0$; in this case the initial data are m_0 and u_0. Let $u^{(0)}$ be the remainder of the division of u_0 by m_0. Then $u^{(0)} \equiv u_0 \bmod m_0$ with $u^{(0)} < m_0$. If $k = 1$ and m_0, m_1 and u_0, u_1 are the data, let $s_0^{(1)} m_0 + s_1^{(0)} m_1 = 1$. We know that a solution is given by $u_1 s_0^{(1)} m_0 + u_0 s_1^{(0)} m_1$. If we take, instead of u_0, the remainder of the division of u_0 by m_0, that is, $u^{(0)}$, then the number $u_1 s_0^{(1)} m_0 + u^{(0)} s_1^{(0)} m_1$ is again a solution, as it is congruent to $u^{(0)}$, hence to u_0, modulo m_0. So, let

$$u^{(1)} = u_1 s_0^{(1)} m_0 + u^{(0)} s_1^{(0)} m_1.$$

Since $s_1^{(0)} m_1 = 1 - s_0^{(1)} m_0$, we may write:

$$u^{(1)} = u_1 s_0^{(1)} m_0 + u^{(0)}(1 - s_0^{(1)} m_0),$$

or $u^{(1)} = u^{(0)} + (u_1 - u^{(0)}) s_0^{(1)} m_0$. This is a solution for this case, as $u^{(1)} \equiv u^{(0)} \equiv u_0 \bmod m_0$, and since $s_0^{(1)} m_0 = 1 - s_1^{(0)} m_1$,

$$u^{(1)} = u^{(0)} + (u_1 - u^{(0)})(1 - s_1^{(0)} m_1),$$

so $u^{(1)} \equiv u_1 \bmod m_1$. Thus, in order to go from $k = 0$ to $k = 1$, we add to the solution $u^{(0)}$ of the case $k = 0$ the difference between the new value u_1 of u and $u^{(0)}$, multiplied by $s_0^{(1)} m_0$. Let now $k = 2$. We have:

$$s_0^{(1)} m_0 + s_1^{(0)} m_1 = 1, \quad s_0^{(2)} m_0 + s_2^{(0)} m_2 = 1, \quad s_1^{(2)} m_1 + s_2^{(1)} m_2 = 1.$$

Set $u^{(2)} = u^{(0)} + (u_1 - u^{(0)}) s_0^{(1)} m_0 + (u_2 - u^{(1)}) s_0^{(2)} m_0 \cdot s_1^{(2)} m_1$. Here we have added to $u^{(1)}$, which is the solution in the case $k = 1$, the difference $u_2 - u^{(1)}$ multiplied by $s_0^{(2)} m_0 \cdot s_1^{(2)} m_1$. It is a solution indeed, since $u^{(2)} \equiv u^{(0)} \equiv u_0 \bmod m_0$ and $u^{(2)} \equiv u^{(1)} \equiv u_1 \bmod m_1$. Moreover, if we write $s_0^{(2)} m_0 = 1 - s_2^{(0)} m_2$ and $s_1^{(2)} m_1 = 1 - s_2^{(1)} m_2$, we have:

$$u^{(2)} = u^{(0)} + (u_1 - u^{(0)}) s_0^{(1)} m_0 + (u_2 - u^{(1)})(1 - s_2^{(0)} m_2)(1 - s_2^{(1)} m_2),$$

which yields $u^{(2)} \equiv u^{(1)} + (u_2 - u^{(1)}) = u_2 \bmod m_2$.

It is now clear how the method works: suppose we know the solution

$$u^{(k-1)} \equiv u_j \bmod m_j,$$

$j = 0, 1, \ldots, k - 1$; then:

$$u^{(k)} = u^{(k-1)} + (u_k - u^{(k-1)}) \prod_{j=0}^{k-1} s_j^{(k)} m_j \tag{1.2}$$

($u^{(0)} = u_0 \bmod m_0$), and the integer we are looking for is $u = u^{(n)}$.

Remarks. 1. It may be useful to split the product $\prod_{j=0}^{k-1} s_j^{(k)} m_j$ and to compute separately the factors $s_k = \prod_{j=0}^{k-1} s_j^{(k)}$ and $q_k = \prod_{j=0}^{k-1} m_j$, for $k = 1, 2, \ldots, n$.

2. Since $s_i^{(k)} m_i + s_k^{(i)} m_k = 1$, we have $s_i^{(k)} m_i \equiv 1 \bmod m_k$, $i \neq k$. It follows that $s_i^{(k)} \equiv m_i^{-1} \bmod m_k$ or $s_i^{(k)}$ is the inverse of m_i modulo m_k.

Set now $a_k = (u_k - u^{(k-1)}) \cdot s_k \bmod m_k$, with $a_0 = u^{(0)} = u_0 \bmod m_0$. Formula (1.2) becomes $u^{(k)} = u^{(k-1)} + a_k q_k$, and we have the following algorithm (in order to simplify computations, it is better, at each step, to reduce the present value of u modulo the following m_k):

input: m_i, u_i, s_i, q_i;
output: u;

$u := u_0$;
for $k := 0$ to $n - 1$ do:
$u := u \bmod m_k$;
$a := (u_k - u)s_{k+1} \bmod m_{k+1}$;
$u : u + a q_{k+1}$.

Example. Let us see how Newton method works for the example seen above. Compute the q_ks:

$$q_1 = m_0 = 7,$$
$$q_2 = m_0 m_1 = 7 \cdot 11 = 77,$$
$$q_3 = m_0 m_1 m_2 = 7 \cdot 11 \cdot 13 = 1001.$$

For the s_ks, we have:

- $k = 1$:

$$s_1 = s_0^{(1)} \equiv m_0^{-1} = 7^{-1} \bmod 11 = 8 \equiv -3 \bmod 11;$$

- $k = 2$:

$$s_0^{(2)} \equiv m_0^{-1} = 7^{-1} \bmod 13 = 2, \quad s_1^{(2)} \equiv m_1^{-1} = 11^{-1} \bmod 13 = 6,$$

so, $s_2 \equiv 2 \cdot 6 \bmod 13 = 12 \bmod 13 = 12$;

- $k = 3$:

$$s_0^{(3)} \equiv m_0^{-1} = 7^{-1} \bmod 15 = 13 \equiv -2 \bmod 13,$$
$$s_1^{(3)} \equiv m_1^{-1} = 11^{-1} \bmod 15 = 11 \equiv -4 \bmod 11,$$
$$s_2^{(3)} \equiv m_2^{-1} = 13^{-1} \bmod 15 = 7,$$

and $s_3 \equiv -2 \cdot -4 \cdot 7 = 56 \bmod 15 = 11 \equiv -4 \bmod 15.$

So the algorithms yields:

- $k = 0$:

$$u := u_0 \bmod 7 = -1 \bmod 7 = 6;$$

- $k = 1$:

$$u := 6,$$
$$a := (-2 - 6) \cdot 8 = -64 \bmod 11 = 2,$$
$$u := 6 + 2 \cdot 7 = 20.$$

Now: $20 \equiv -1 \bmod 7$ and $\equiv -2 \bmod 11$;

- $k = 2$:

$$u := 20 \bmod 13 = 7,$$
$$a := (2 - 7) \cdot 2 = -10 \bmod 15 = 5,$$
$$u := 20 + 5 \cdot 77 = 405.$$

Here $405 \equiv -1 \bmod 7$, $\equiv -2 \bmod 11$, and $\equiv 2 \bmod 13$;

- $k = 3$:

$$u := 405 \bmod 15 = 0,$$
$$a := (6 - 0) \cdot 11 = 66 \bmod 15 = 6,$$
$$u := 405 + 6 \cdot 1001 = 6411,$$

which is the integer we found by Lagrange method. Note now that:

$$u = u^{(n)} = (u^{(0)} - u^{(-1)}) + (u^{(1)} - u^{(0)}) + \ldots + (u^{(n)} - u^{(n-1)}).$$

Setting $u^{(-1)} = 0$ we get:

$$u = a_0 + a_1 m_0 + a_2 m_0 m_1 + \ldots + a_n m_0 m_1 \cdots m_{n-1}. \tag{1.3}$$

The integers a_i provide what is called the *Newtonian representation* of u in the "mixed basis" $1, m_0, m_0 m_1, \ldots, m_0 m_1 \cdots m_{n-1}$:

$$u = \langle a_0, a_1, \ldots, a_n \rangle.$$

In our example we have $6411 = 6 + 2 \cdot 7 + 5 \cdot 77 + 6 \cdot 1001$, hence:

$$6411 = \langle 6, 2, 5, 6 \rangle.$$

1.3 Polynomials

As for the integers, there is a Euclidean division for polynomials with coefficients in a field: if $f = f(x)$ and $g = g(x)$ are two polynomials, there is a (unique) pair of polynomials $q = q(x)$ and $r = r(x)$, with $\partial r < \partial g$, such that:

$$f = gq + r.$$

(The symbol ∂ denotes the degree of the polynomial. Keep in mind that the elements of the field – the *constants* – have degree 0, except 0 – the *zero polynomial* – for which the degree is not defined.) Let us see how to find q and r. The initial values for q and r are given by:

$$q = 0 \text{ and } r = f;$$

indeed:

$$f = 0 \cdot g + f,$$

but the condition $\partial r < \partial g$ is in general not satisfied. (Note that $q = 0$ and $r = f$ is the solution when $\partial g > \partial f$.) If $r \neq 0$ and $\partial r \geq \partial g$, consider the leading (i.e., largest degree) monomials of r and g - let them be $m(r)$ and $m(g)$ - and their quotient $m(r)/m(g)$. Take as new values of q and r:

$$q + \frac{m(r)}{m(g)} \text{ and } r - \frac{m(r)}{m(g)}g.$$

When $r = 0$ or $\partial r < \partial g$, the procedure stops. So we have the following algorithm:

input: f, g;
output: q, r;

$q := 0$;
$r := f$;
while $r \neq 0$ and $\partial(r) > \partial(g)$ do:
$q := q + \frac{m(r)}{m(g)}$;
$r := r - \frac{m(r)}{m(g)}g$.

Example.
input: $f = x^4 + x^2 - 2x + 1$, $g = x^2 + 1$;
output: q, r;

$q := 0$;
$r := f$.
$r \neq 0$? Yes;
$\partial r > \partial g$? Yes.
$\frac{m(r)}{m(g)} = \frac{x^4}{x^2} = x^2$;
$q := 0 + x^2$;
$r := f - x^2 g = -2x + 1$.
$\partial r > \partial g$? No.

The algorithm stops, giving as its output the present values of q and r, that is, x^2 and $-2x + 1$, which are the quotient and the remainder of the division.

Let us see now why the algorithm works, that is, why it actually does what we require (finding q and r with $f = qg + r$ and $0 \leq \partial r < \partial g$). Firstly, because $f = qg + r$ is true for the starting values 0 and f of q and r, and then because this equality remains true when we substitute the new values for q and r:

$$f = qg + r = (q + \frac{m(r)}{m(g)})g + (r - \frac{m(r)}{m(g)}g).$$ (1.4)

Finally, because either the final value of r is $r = 0$ or it satisfies $\partial r < \partial g$ (and the algorithm stops). Indeed, the degree of $r - \frac{m(r)}{m(g)}g$ is less than the degree of r; let

$$r = a_0 x^m + a_1 x^{m-1} + \cdots + a_m,$$
$$g = b_0 x^k + b_1 x^{k-1} + \cdots + b_k,$$

and let $m \geq k$. Then:

$$r - \frac{m(r)}{m(g)}g = (a_0 x^m + a_1 x^{m-1} + \cdots) - \frac{a_0}{b_0} x^{m-k}(b_0 x^k + b_1 x^{k-1} + \cdots)$$
$$= a_0 x^m + a_1 x^{m-1} + \cdots + a_m - a_0 x^m - \cdots$$
$$= a_1 x^{m-1} + \cdots$$

while the remaining terms have degree less than $m - 1$ (or the whole polynomial is zero).

Remark. Of course, in (1.4) the equality holds no matter which polynomial is substituted for $m(r)/m(g)$. However, this particular polynomial is one that lowers the degree.

Given a nonzero polynomial g, consider the polynomial $f = gq + r$, with r either of degree less than the degree of g or equal to zero. Then f has degree $\partial g + \partial q$, so the equality $gq + r = 0$ implies $q = r = 0$. Let us show that, given f, g, with $g \neq 0$, there exist q and r such that $f = gq + r$, with $\partial r < \partial g$ or $r = 0$. If $\partial f = n$, this equality, with unknown q and r, gives rise to a system of $n + 1$ linear equations in the $n + 1$ unknowns given by the coefficients of q and r. The determinant D of this system is not zero. Indeed, if $D = 0$, the homogeneous system obtained by setting $gq + r = 0$ would have a nonzero solution (it would be satisfied with some nonzero coefficient of q or r), and this contradicts the remark above. So $D \neq 0$, and by Cramer's rule the solution exists and is unique.

Example. Let $f = x^4 + 2x^3 - 1$ and $g = x^2 + 1$. Since f has degree 4, q has to have degree 2, $q = ax^2 + bx + c$, and $\partial r < \partial g$, $r = dx + e$ or $r = 0$. So, the expression $f = gq + r$ can be written as $x^4 + 2x^3 - 1 = (x^2 + 1)(ax^2 + bx + c) + dx + e = ax^4 + bx^3 + (a + c)x^2 + (b + d)x + c + e$. By equating the coefficients of the two sides we get the following system of 5 (= 4 + 1) equations in the 5 unknowns a, b, c, d, e:

$$a = 1, \ b = 2, \ a + c = 0, \ b + d = 0, \ c + e = -1,$$

admitting the solution

$$a = 1, \ b = 2, \ c = -1, \ d = -2, \ e = 0,$$

hence $q = x^2 + 2x - 1$ and $r = -2x$. Note that the matrix of the system

$$\begin{pmatrix} 1 & 0 & 0 & 0 & 0 \\ 0 & 1 & 0 & 0 & 0 \\ 1 & 0 & 1 & 0 & 0 \\ 0 & 1 & 0 & 1 & 0 \\ 0 & 0 & 1 & 0 & 1 \end{pmatrix}$$

has nonzero determinant (equal to 1).

If $r = 0$ we say that g *divides* f or that f is a *multiple of* g. If a polynomial has no divisors except itself and the constants, it is said to be *irreducible*. Otherwise, that is, when it is a product of two polynomials, both of lesser degree, the polynomial is said to be *reducible*.

Corollary 1.1. *The remainder of the division of a polynomial $f(x)$ by $x - a$, where a is a constant, is $f(a)$, the value of $f(x)$ at a. In particular, if a is a root of the polynomial $f(x)$, then this remainder is zero, that is, $f(x)$ is a multiple of $x - a$.*

Proof. The division of $f(x)$ by $x - a$ yields $f(x) = (x - a)q(x) + r(x)$, with $\partial r(x) < \partial(x - a) = 1$, so either $r(x)$ is zero everywhere, or $\partial r(x) = 0$, that is, $r(x) = c$, a constant. In the first case $f(a) = 0$, that is, a is a root of $f(x)$; in the second one $f(a) = c$. If $c = 0$, we have again that $r(x)$ is the zero polynomial (since it is constant, if it takes the value 0 once, is zero everywhere) and a is a root.   ◇

If $f(x) = (x - a)q(x)$ and $q(a) = 0$, then a is (at least) a *double* root of $f(x)$; in this case we have that $(x - a)^2$ divides $f(x)$, and a is a root of *multiplicity (at least) m* if $(x - a)^m$ divides $f(x)$.

In general, if $f(x) = g(x)q(x) + r(x)$, the value of $f(x)$ computed at a root a of $g(x)$ is equal to the remainder $r(x)$ computed at a, $f(a) = r(a)$, the advantage being that the degree of $r(x)$ is in general much smaller than the degree of $f(x)$. For instance, to compute $f(x) = x^3 - 2x + 1$ at $\sqrt{2}$, divide $f(x)$ by $x^2 - 2$: we find 1 as remainder, and this is the value of $f(\sqrt{2})$ we were looking for. In Corollary 1.1 we consider the special case when $g(x) = x - a$, which has a as root.

Corollary 1.2. *A polynomial $f(x)$ of degree $n \geq 1$ has at most n roots (in the field or in an extension of the field).*

Proof. By induction on n. If $n = 1$, then $f(x) = ax + b$ has the unique root $x = -b/a$. Let $\partial f > 1$. If $f(x)$ has no roots, there is nothing to prove. Otherwise, let a be a root. By the previous corollary, $f(x)$ is a multiple of $x - a$: $f(x) = (x - a)q(x)$. Now, if $b \neq a$ is a root of $f(x)$, it is a root of $q(x)$ as well, since $0 = f(b) = (b - a)q(b)$, and as $b - a \neq 0$ we have $q(b) = 0$. The polynomial $q(x)$ has degree $n - 1$, so, by induction, has at most $n - 1$ roots. These are roots of $f(x)$ too; adding the root a, $f(x)$ has at most n roots.   ◇

If we keep in mind the fundamental theorem of algebra: *a polynomial of degree $n \geq 1$ with coefficients in the complex field C has at least a root in C,* we have:

Corollary 1.3. *A polynomial of degree $n \geq 1$ with coefficients in C has exactly n roots in C (counting each root with its multiplicity).*

Corollary 1.4. *If two polynomials $f(x)$ and $g(x)$ of degree at most n take the same values for $n + 1$ distinct values of the variable, then they coincide: $f(x) = g(x)$.*

Proof. The polynomial $f(x) - g(x)$ has degree at most n and admits $n + 1$ roots. It cannot have positive degree, or it would have, by the previous corollary, at most n roots, and it cannot have degree zero, or it would be a nonzero constant, so it would have no roots. Then it has to be the zero polynomial: $f(x) - g(x) = 0$, and hence $f(x) = g(x)$. ◇

In particular if, given $n + 1$ distinct values $x_0, x_1, \ldots, x_n$ of the variable, there is a polynomial of degree at most n that takes $n + 1$ given values on the x_ks, then this polynomial is unique. Its existence will be discussed in next section.

Just like for integers, we may define the *greatest common divisor* of two polynomials $f(x)$ and $g(x)$: it is *the polynomial $d(x)$ that has as divisors exactly all common divisors of $f(x)$ and $g(x)$.* (For polynomials the divisors are defined up to multiplicative constants.) The proof of the existence of $d(x)$ is constructive. As for the integers, it relies on the Euclidean algorithm, which has now the following form: dividing $f = f(x)$ by $g = g(x)$ we have:

$$f = gq + r, \qquad \partial r < \partial g,$$

and when we go on dividing, we have (setting $r = r_1$ and $q = q_1$):

$$g = r_1 q_2 + r_2, \qquad \partial r_2 < \partial r_1,$$
$$r_1 = r_2 q_3 + r_3, \qquad \partial r_3 < \partial r_2,$$
$$\vdots$$
$$r_{k-3} = r_{k-2} q_{k-1} + r_{k-1}, \qquad \partial r_{k-1} < \partial r_{k-2},$$
$$r_{k-2} = r_{k-1} q_k,$$

and $r_k = 0$. The last nonzero remainder r_{k-1} is the greatest common divisor $d = (f, g)$ of the two polynomials. If $r = 0$, then $(f, g) = g$, and if $d(x)$ is a constant $f(x)$ and $g(x)$ are *relatively prime.*

The existence of a greatest common divisor allows us to extend to polynomials some of the results that hold for integers. In particular, we have that *every polynomial can be decomposed as a product of irreducible polynomials,* and this happens, up to reordering the factors and multiplying by a constant, *in a unique way.*

We have already seen (Corollary 1.1) that if a polynomial has a root, then it can be reduced. The converse is not true in general. For instance, on the real field the polynomial $x^4 + 2x^2 + 1$ can be written as $(x^2 + 1)(x^2 + 1)$, but has no real roots.

Examples. **1.** (*Derivative of a polynomial*) The existence of Euclidean division allows us to define the derivative of a polynomial in a purely algebraic way. Indeed, let

$$f(x) = a_0 x^n + a_1 x^{n-1} + \cdots + a_n$$

be a polynomial over a field, and α an element of the field or of an extension of the field. Then, dividing $f(x)$ by $x - \alpha$ we have:

$$f(x) = (x - \alpha)q(x) + r(x), \tag{1.5}$$

with $r(x)$ a constant. Performing explicitly the division, we find, for the quotient:

$$q(x) = a_0 x^{n-1} + (a_0\alpha + a_1)x^{n-2} + (a_0\alpha^2 + a_1\alpha + a_2)x^{n-3} + \cdots$$
$$+ a_0\alpha^{n-1} + a_1\alpha^{n-2} + \cdots + a_{n-2}\alpha + a_{n-1},$$

and for the remainder:

$$r(x) = a_0\alpha^n + a_1\alpha^{n-1} + \cdots + a_{n-1}\alpha + a_n.$$

(Note that $r(x)$ is the value $f(x)$ takes in α, so it is 0 if α is a root of $f(x)$, and conversely.) So the coefficients of $q(x)$ are polynomials in α of degree 0, 1, ..., $n-1$, where the k-th polynomial has as its coefficients the first k coefficients of $f(x)$. Computing $q(x)$ in α we find $n - k$ times the term $a_k\alpha^{n-k-1}$, $k = 0, 1, \ldots, n-1$, and hence:

$$q(\alpha) = na_0\alpha^{n-1} + (n-1)a_1\alpha^{n-2} + 2a_{n-2}\alpha + a_{n-1}.$$

If we write the right-hand polynomial as a polynomial in x, we get the expression of what we call the *derivative* of the polynomial $f(x)$. (The derivative of (1.5) computed as taught by calculus is:

$$f'(x) = q(x) + (x - \alpha)q'(x),$$

so $f'(\alpha) = q(\alpha)$, as above.) If $f(\alpha) = 0$ and $f'(\alpha) = 0$, then $q(\alpha) = 0$, so $q(x) = (x - \alpha)q_1(x)$. Hence, from (1.5) with $r(\alpha) = 0$, we have $f(x) = (x - \alpha)^2 q_1(x)$ and α is (at least) a double root of $f(x)$. Conversely, if α is (at least) a double root of $f(x)$, from (1.5) with $r(x) = 0$ we get $q(\alpha) = 0$, so $f'(\alpha) = 0$ too. It follows that a root of a polynomial is a multiple root if and only if it is a root of the derivative of the polynomial too.

2. (*Horner's method*) From (1.5) we see that the value $f(\alpha)$ of a polynomial $f(x)$ at a point α is the value at α of the remainder of the division of $f(x)$ by $x - \alpha$. This division yields a way to compute $f(\alpha)$ which is more efficient that the usual method involving the computation of the powers of α. With the usual method we have to perform the $n - 1$ multiplications $\alpha \cdot \alpha = \alpha^2$, $\alpha^2 \cdot \alpha = \alpha^3, \ldots, \alpha^{n-1} \cdot \alpha = \alpha^n$, and then the n multiplications $a_{n-i}\alpha^i$; all in all, $2n - 1$ multiplications. Note how we find instead the coefficients of the quotient $q(x)$ from those of $f(x)$: starting from a_0, we multiply by α and add the following coefficient:

$$a_0, a_0\alpha + a_1, (a_0\alpha + a_1)\alpha + a_2, \ldots$$

The last coefficient of $q(x)$ is a polynomial of degree $n - 1$ in α; multiplying again by α and adding a_n we also get the remainder of the division, that is, the value of $f(x)$ in α:

$$f(\alpha) = ((\ldots (a_0\alpha + a_1)\alpha + a_2)\alpha + \cdots) + a_n.$$

The last equality gives a method (*Horner method*) to compute a polynomial at a point α that requires n multiplications, that is, about half of those required by the usual method, and n additions.

Just like for integers, we have, with the same proof, a Bézout identity for polynomials:

Theorem 1.4 (Bézout's identity). *Let $d(x) = (f(x), g(x))$. Then there are two polynomials $a(x)$ and $b(x)$ such that:*

$$d(x) = a(x)f(x) + b(x)g(x).$$

Remark. In particular, if $f(x)$ and $g(x)$ are relatively prime, then we have the above identity with $d(x) = d$, a constant. Dividing $a(x)$ and $b(x)$ by d, we have two polynomials $a'(x)$ and $b'(x)$ such that $a'(x)f(x) + b'(x)g(x) = 1$.

The Chinese remainder theorem, which relies on Bézout identity, holds for polynomials too, with an analogous proof.

Theorem 1.5 (Chinese remainder theorem for polynomials). *Let*

$$m_0(x), m_1(x), \ldots, m_n(x),$$

be pairwise coprime polynomials, and let

$$u_0(x), u_1(x), \ldots, u_n(x)$$

be arbitrary polynomials. Then there exists a polynomial $u(x)$ such that

$$u(x) \equiv u_k(x) \bmod m_k(x), \ k = 0, 1, \ldots, n,$$

and this $u(x)$ is unique modulo the product $m(x) = m_0(x)m_1(x) \cdots m_n(x)$. (That is, $u(x)$ is the only polynomial of degree less than the degree of $m(x)$ that solves the congruence system.)

Let us prove a lemma first.

Lemma 1.1. *Let f, g and h be three polynomials with $\partial h < \partial f + \partial g$, and suppose that, for suitable polynomials a and b, the relation*

$$af + bg = h$$

holds. Then the same relation holds for two polynomials a_1 and b_1 such that $\partial a_1 < \partial g$ and $\partial b_1 < \partial f$.

Proof. First of all, note that if $\partial a < \partial g$, then necessarily $\partial b < \partial f$, otherwise $\partial h = \partial(af + bg) = \partial bg \geq \partial f + \partial g$, against the hypothesis. Suppose now that $\partial a \geq \partial g$; dividing we get $a = gq + r$ with $\partial r < \partial g$; so, set $a_1 = a - gq$ and $b_1 = b + fq$ and we get $a_1 f + b_1 g = h$; since $\partial a_1 < \partial g$ we find $\partial b_1 < \partial f$, by the above argument.   ◇

Corollary 1.5. *Let f and g be two relatively prime polynomials. Then there exist two polynomials a and b with $\partial a < \partial g$ and $\partial b < \partial f$, and such that*

$$af + bg = 1.$$

Moreover, a and b are uniquely determined.

Proof. Two polynomials a and b such that $af + bg = 1$ exist, since f and g are relatively prime. So the claim follows from the lemma, if we take as h the constant polynomial 1. If a_1, b_1 is another such pair, then by subtracting the relation $a_1 f + b_1 g = 1$ from the former, we get $(a - a_1)f = (b_1 - b)g$, and since $(f, g) = 1$, g has to divide $a - a_1$; but $\partial(a - a_1) < \partial g$, and hence $a - a_1 = 0$ and $a_1 = a$, so we have $b_1 = b$ too.   ◇

We can now prove Theorem 1.5.

Proof. Let $m(x) = \prod_{k=0}^{n} m_k(x)$ and let $l_k(x) = \frac{m(x)}{m_k(x)}$. Then $(m_k(x), l_k(x)) = 1$, and by the corollary there are, uniquely, $a_k(x)$ and $b_k(x)$ such that $\partial a_k(x) < \partial l_k(x)$ and $\partial b_k(x) < \partial m_k(x)$, and

$$a_k(x)m_k(x) + b_k(x)l_k(x) = 1.$$

Set $L_k(x) = b_k(x)l_k(x)$, and note that since $\partial b_k(x) < \partial m_k(x)$, the degree of L_k is less than the degree of $m(x)$. Moreover, by construction,

$$L_k(x) \equiv 1 \mod m_k(x),$$
$$L_k(x) \equiv 0 \mod m_i(x),\ i \neq k.$$

Thus,

$$u(x) = \sum_{k=0}^{n} u_k(x)L_k(x)$$

is the required polynomial.   ◇

In the ring $A = K[x]/(m(x))$ of the polynomials with coefficients in the field K and of degree less than the degree of $m(x)$, endowed with the usual sum and the product followed by the reduction mod $m(x)$, the polynomials $L_k(x)$ are orthogonal idempotents, summing to 1:

1. $L_0(x) + L_1(x) + \ldots + L_n(x) \equiv 1 \bmod m(x)$.

   This follows from the fact that $\sum L_k(x) - 1$ is divisible by all the $m_k(x)$s, and hence by $m(x)$.

2. $L_i(x)L_j(x) \equiv 0 \bmod m(x), i \neq j$.

   The factor $m_i(x)$ that is missing in $L_i(x)$ appears in $L_j(x)$. In the product $L_i(x)L_j(x)$ all the factors $m_k(x)$ appear, so $m(x)$ is a factor of this product.

3. $L_k(x)^2 \equiv L_k(x) \bmod m(x)$.

   This follows by multiplying 1. by $L_k(x)$ and then keeping in mind 2.

Every element $f(x) \in A$ also has a unique decomposition of the form:

$$f(x) \equiv f_0(x)L_0(x) + f_1(x)L_1(x) + \cdots + f_n(x)L_n(x) \bmod m(x),$$

where $f_i(x)$ is the remainder of the division of $f(x)$ by $m_i(x)$. Indeed, consider $f(x)L_k(x)$. If $\partial f(x) \geq \partial m_o(x)$, let $f(x) = m_o(x)q(x) + r(x)$, so $f(x)L_k(x) = m_0(x)q(x)L_k(x) + r(x)L_k(x)$. But $m_0(x)L_k(x) \equiv 0 \bmod m(x)$, and hence $f(x)L_k(x) \equiv f_k(x)L_k(x) \bmod m(x)$ with $f_k(x) = r(x)$. As in the case of integers, the properties above allows us to decompose the ring A as a direct sum of subrings $A_k = \{f(x)L_k(x) \bmod m(x)\}$. Finally, with the same proof as for the integers, we have $A_k \simeq K[x]/(m_k(x))$.

Remark. The ring A is also a vector space over K, as is easily seen; so it is an *algebra*.

Example. Let

$$m_0(x) = x - 1, \quad m_1(x) = x^2 + 1, \quad m_2(x) = x^2 - 2,$$

and $m(x) = m_0(x)m_1(x)m_2(x) = x^5 - x^4 - x^3 + x^2 - 2x + 2$.
Compute $L_0(x)$. With $m_0(x) = x - 1$ and $m_1(x)m_2(x) = x^4 - x^2 - 2$, we get

$$\frac{-x^3 - x^2}{2} \cdot (x - 1) - \frac{1}{2}(x^4 - x^2 - 2) = 1,$$

and hence $L_0(x) = -\frac{1}{2}(x^4 - x^2 - 2)$.

For $L_1(x)$, with $m_1(x) = x^2 + 1$, $m_0(x)m_2(x) = x^3 - x^2 - 2x + 2$ we find

$$\frac{4 - x^2}{6} \cdot (x^2 + 1) + \frac{x + 1}{6}(x^3 - x^2 - 2x + 2) = 1,$$

and $L_1(x) = \frac{x+1}{6}(x^3 - x^2 - 2x + 2) = \frac{1}{6}(x^4 - 3x^2 + 2)$.

Finally, with $m_2(x) = x^2 - 2$, $m_0(x)m_1(x) = x^3 - x^2 + x - 1$ we have

$$-\frac{2+x^2}{3} \cdot (x^2 - 2) + \frac{x+1}{3} \cdot (x^3 - x^2 + x - 1) = 1$$

and $L_2(x) = \frac{x+1}{3} \cdot (x^3 - x^2 + x - 1) = \frac{1}{3}(x^4 - 1)$.

It is immediate to verify that $L_0(x) + L_1(x) + L_2(x) = 1$, as well as the other two properties of the $L_k(x)$s.

Let $x^3 - 1$ be a polynomial in A; we have:

$$x^3 - 1 = (x - 1)q_0(x) + 0,$$
$$x^3 - 1 = (x^2 + 1)q_1(x) + (-x - 1),$$
$$x^3 - 1 = (x^2 - 2)q_2(x) + 2x - 1.$$

It follows that

$$x^3 - 1 \equiv 0 \cdot L_0(x) + (-x - 1)L_1(x) + (2x - 1)L_2(x) \mod m(x).$$

Let us verify this. The sum in the right-hand side equals $\frac{1}{2}(x^5 - x^4 + x^3 + x^2 - 2x + 2)$ and, divided by $m(x)$, gives a remainder equal to $\frac{1}{2}(2x^3 - 2)$, that is, $x^3 - 1$.

A particular but very important case of the Chinese remainder theorem is that in which the polynomials $m_k(x)$ are of first degree and of the form $x - x_k$, and the $u_k(x)$s are constants. It is the case we shall discuss in detail in next section.

Exercises

10. State and solve the exercise from 1 to 5 and 7 for polynomials.

11. Prove that if $f = \prod_i f_i$ with $(f_i, f_j) = 1$, and a is an arbitrary polynomial, we have:

$$\frac{a}{f} = g + \frac{a_1}{f_1} + \frac{a_2}{f_2} + \cdots + \frac{a_n}{f_n}$$

where g is a polynomial and $\partial a_i < \partial f_i$, and the a_is are unique under these conditions. Moreover, if $\partial a < \partial f$, then $g = 0$. (This decomposition of a/f is called the *simple fraction decomposition*.)

12. A polynomial of degree ≤ 3 over an arbitrary field is reducible if and only if it admits a root.

13. Give an example demonstrating the fact that, if the coefficients are not in a field, the uniqueness of the decomposition into irreducible polynomials may fail.

14. Prove that in the complex field two polynomials are not relatively prime if and only if they have a common root.

15. Prove that the gcd of two polynomials $f(x)$ and $g(x)$ does not depend on the field of coefficients.

16. Prove that the polynomial $f(x) - x$ divides $f(f(x)) - x$. [*Hint*: having set $f(x) = y$, the remainder of the division of $f(y) - x$ by $y - x$ is the value of $f(y) - x$ at the point x.]

17. Using the data from the example above, determine the expression of the polynomial $x^3 + 1$ as a combination of the L_ks.

1.4 Polynomial interpolation

The classic problem of polynomial interpolation has the following form: given $n + 1$ distinct numbers ($n + 1$ elements of a field) $x_0, x_1, \ldots, x_n$, and $n + 1$ arbitrary numbers $u_0, u_1, \ldots, u_n$, find a polynomial $u(x)$ of degree at most n whose value at x_k is u_k, $k = 0, 1, \ldots, n$ (we know, by Corollary 1.4, that such a polynomial, if it exists, is unique).

This problem is a particular case of that solved by the Chinese remainder theorem. Indeed, if $u(x)$ is such that $u(x_k) = u_k$, then the polynomial $u(x) - u_k$ has the root x_k, so it is divisible by $x - x_k$; in other words, $u(x) - u_k \equiv 0 \bmod (x - x_k)$, or:

$$u(x) \equiv u_k \bmod (x - x_k), \tag{1.6}$$

$k = 0, 1, \ldots, n$. Since the numbers x_k are pairwise distinct, the polynomials $x - x_k$ are pairwise coprime. So the hypotheses of the Chinese remainder theorem are satisfied, with $m_k(x) = x - x_k$ and the polynomials $u_k(x) = u_k$ of degree zero (constants). So the theorem ensures the existence of a polynomial $u(x)$ such that (1.6) holds, and its uniqueness modulo the product $\prod_{i=0}^{n}(x - x_k)$. Since this product has degree $n + 1$, $u(x)$ is the only polynomial of degree at most n (or, possibly, the zero polynomial) that satisfies the conditions we required. It is called the *Lagrange (interpolating) polynomial*.

Remark. The existence and uniqueness of such a polynomial $u(x)$ can be seen using the theory of systems of linear equations. For the unknown polynomial $u(x) = \sum_{i=0}^{n} a_i x^i$, the relations $\sum_{i=0}^{n} a_i x_k^i = u_k$, $k = 0, 1, \ldots, n$ have to hold. So we have a system of $n + 1$ equations in the $n + 1$ unknowns $a_0, a_1, \ldots, a_n$, whose determinant is the Vandermonde polynomial V of x_i. Since the x_i are pairwise distinct, we have $V \neq 0$, so the solution exists and is unique.

As in the case of integers, we have two methods to compute $u(x)$.

1.4.1 Lagrange's method

Determine the polynomials $L_k(x)$, defined in the previous section, with the same procedure as for integers (cf. 1.2.1). In this case, they are called *Lagrange polynomials*, and their computation is made easy by the fact that the $m_k(x)$s are now linear. Indeed, given $x - x_i$ and $x - x_j$, it is immediate to find s_i^j and

s_j^i such that $s_i^j(x - x_i) + s_j^i(x - x_j) = 1$: it suffices to take the two constants

$$s_i^j = \frac{1}{x_j - x_i}, \quad s_j^i = \frac{1}{x_i - x_j}.$$

We get:

$$\frac{1}{x_j - x_i}(x - x_i) + \frac{1}{x_i - x_j}(x - x_j) = \frac{x - x_i - (x - x_j)}{x_j - x_i} = \frac{x_j - x_i}{x_j - x_i} = 1.$$

From this follows:

$$L_k(x) = \prod_{i \neq k} s_i^{(k)} \cdot \prod_{i \neq k}(x - x_i) = \frac{\prod_{i \neq k}(x - x_i)}{\prod_{i \neq k}(x_k - x_i)}.$$

So all the $L_k(x)$s are of degree n, and the polynomial $u(x)$ we are looking for is

$$u(x) = \sum_{k=0}^{n} u_k L_k(x).$$

It is true for this polynomial that $u(x_k) = u_k$, since $L_k(x_k) = 1$ and $L_k(x_j) = 0$, $j \neq k$, and this is Lagrange interpolating polynomial.

To compute $u(x)$ by this method we have to perform $2(n - 1)$ multiplications for each of the $L_k(x)$ ($n - 1$ for the numerator, and as many for the denominator), so computing all the $L_k(x)$s requires $2(n - 1)(n + 1) = 2n^2 - 2$ multiplications. We have to add then the $n + 1$ multiplications for the u_ks; all in all, we need $2n^2 + n - 1$ multiplications.

Example. Let $x_0 = 1$, $x_1 = 4$, $x_2 = 6$, $x_3 = 11$, and $u_0 = 10$, $u_1 = 334$, $u_2 = 1040$, $u_3 = 5920$. Thus:

$$L_0(x) = \frac{(x - 4)(x - 6)(x - 11)}{(1 - 4)(1 - 6)(1 - 11)} = \frac{x^3 - 21x^2 + 134x - 264}{-150},$$

$$L_1(x) = \frac{(x - 1)(x - 6)(x - 11)}{(4 - 1)(4 - 6)(4 - 11)} = \frac{x^3 - 18x^2 + 83x - 66}{42},$$

$$L_2(x) = \frac{(x - 1)(x - 4)(x - 11)}{(6 - 1)(6 - 4)(6 - 11)} = \frac{x^3 - 16x^2 + 59x - 44}{-50},$$

$$L_3(x) = \frac{(x - 1)(x - 4)(x - 6)}{(11 - 1)(11 - 4)(11 - 6)} = \frac{x^3 - 11x^2 + 34x - 24}{350}.$$

From this follows:

$$u(x) = 10L_0(x) + 334L_1(x) + 1040L_2(x) + 5920L_3(x) = 4x^3 + 5x^2 - x + 2.$$

Let $m(x) = (x - x_0)(x - x_1) \cdots (x - x_n)$. We know that the $L_k(x)$s are orthogonal idempotents summing to 1 mod $m(x)$; but here their sum is 1, not just 1 mod $m(x)$:

$$L_0(x) + L_1(x) + \cdots + L_n(x) = 1.$$

This can be proved either by computing the sum, or remarking that the left-hand sum is a polynomial of degree $\leq n$ that has value 1 for the $n+1$ values $x_0, x_1, \ldots, x_n$ of the variable x, so it equals 1 everywhere.

If we write:

$$f(x) \equiv f_0(x)L_0(x) + f_1(x)L_1(x) + \cdots + f_n(x)L_n(x) \text{ mod } m(x),$$

which corresponds to the decomposition of the algebra $A = K[x]/(m(x))$ of polynomials of degree at most n, the $f_k(x)$s are now the remainders of the division of $f(x)$ by $x - x_k$, so $f_k(x) = f(x_k)$; so, the polynomials $f_k(x)$ are constant, and the subrings $A_k \simeq K[x]/(m_k(x))$ are all isomorphic to K:

$$K[x]/(m(x)) = K \oplus K \oplus \cdots \oplus K.$$

The standard basis of this vector space of dimension $n+1$ is given by the monomials $1, x, x^2, \ldots, x^n$. we shall prove now that the $L_k(x)$s form a basis too.

Theorem 1.6. *The polynomials $L_k(x)$ form a basis of the vector space of polynomials of degree at most n.*

Proof. Let $f(x)$ be a polynomial of degree at most n. Then:

$$f(x_0)L_0(x) + f(x_1)L_1(x) + \cdots + f(x_n)L_n(x)$$

is a polynomial of degree at most n that has the same values as $f(x)$ on the $n+1$ points $x_0, x_1, \ldots, x_n$, so it coincides with $f(x)$. Hence, $f(x)$ is a linear combination of the $L_k(x)$s (and the coefficients are the $f(x_k)$); this proves that the $L_k(x)$s span our space. Moreover, the expression we have written is the unique linear combination of the $L_k(x)$s that yields $f(x)$, because if

$$f(x) = a_0 L_0(x) + a_1 L_1(x) + \cdots + a_n L_n(x),$$

then, by successively computing $x_0, x_1, \ldots, x_n$, we find $a_k = f(x_k)$. So the $L_k(x)$s are linearly independent. $\diamond$

Remark. Unlike the standard basis, all the elements of this basis have the same degree n. So every choice of $n+1$ distinct numbers $x_0, x_1, \ldots, x_n$ determines a basis of the space all whose elements have the same degree n.

Let us see now how the change of basis works from a basis consisting of the $L_k(x)$s (for some choice of the x_is), to the standard basis $1, x, x^2, \ldots, x^n$. For a monomial x^i we have:

$$x^i = x_0^i L_0(x) + x_1^i L_1(x) + \ldots + x_n^i L_n(x).$$

From this it follows that the change of basis matrix from the basis $(L_k(x))$ to the basis (x^i) is the $(n+1) \times (n+1)$ Vandermonde matrix:

$$V = \begin{pmatrix} 1 & 1 & \cdots & 1 \\ x_0 & x_1 & \cdots & x_n \\ x_0^2 & x_1^2 & \cdots & x_n^2 \\ \vdots & \vdots & \ddots & \vdots \\ x_0^n & x_1^n & \cdots & x_n^n \end{pmatrix}.$$

(Since this is a change of basis matrix, V is non-singular. So we have a proof of the fact that the Vandermonde determinant is not zero.) The inverse matrix V^{-1} of V is the change of basis matrix from (x^i) to $(L_k(x))$. So the entries of V^{-1} allows us to write the polynomials $L_k(x)$ in terms of the powers of x. But this just means writing the $L_k(x)$s in the usual form as polynomials in x. From this follows that V^{-1} is the matrix in which the entries in the k-th rows are the coefficients of the k-th Lagrange polynomial $L_k(x)$, $k = 0, 1, \ldots, n$.

Example. As in the previous example, take

$$x_0 = 1, x_1 = 4, x_2 = 6, x_3 = 11.$$

In this case the Vandermonde matrix is:

$$V = \begin{pmatrix} 1 & 1 & 1 & 1 \\ 1 & 4 & 6 & 11 \\ 1 & 16 & 36 & 121 \\ 1 & 64 & 216 & 1331 \end{pmatrix}.$$

The polynomial $L_0(x)$ is

$$L_0(x) = \frac{264}{150} - \frac{134}{150}x + \frac{21}{150}x^2 - \frac{1}{150}x^3,$$

and analogously for the other ones. The matrix V^{-1} is

$$V^{-1} = \begin{pmatrix} \dfrac{264}{150} & -\dfrac{134}{150} & \dfrac{21}{150} & \dfrac{1}{150} \\[2mm] -\dfrac{66}{42} & \dfrac{83}{42} & -\dfrac{18}{42} & \dfrac{1}{42} \\[2mm] \dfrac{44}{50} & -\dfrac{59}{50} & \dfrac{16}{50} & -\dfrac{1}{50} \\[2mm] -\dfrac{24}{350} & \dfrac{34}{350} & -\dfrac{11}{350} & \dfrac{1}{350} \end{pmatrix}.$$

Let us see now another expression for the $L_k(x)$s. Let:

$$m(x) = (x - x_0)(x - x_1) \cdots (x - x_n).$$

We have:

$$L_k(x) = \frac{\prod_{i \neq k}(x - x_i)}{\prod_{i \neq k}(x_k - x_i)} = \frac{1}{\prod_{i \neq k}(x_k - x_i)} \cdot \frac{m(x)}{x - x_k},$$

and note that $\prod_{i \neq k}(x_k - x_i)$ is $\frac{m(x)}{x - x_k}$ computed in x_k. Then, having set $\frac{m(x)}{x - x_k} = q_k(x)$, we have $L_k(x) = \frac{q_k(x)}{q_k(x_k)}$. Differentiating $m(x) = (x - x_k)q_k(x)$ we get $m'(x) = q_k(x) + (x - x_k)q_k'(x)$, hence $q_k(x_k) = m'(x_k)$, and finally:

$$L_k(x) = \frac{q_k(x)}{m'(x_k)} = \frac{m(x)}{m'(x_k)(x - x_k)} = \frac{1}{m'(x_k)} \cdot \frac{m(x)}{x - x_k}$$

$$= \frac{1}{m'(x_k)}(x - x_0) \cdots (x - x_{k-1})(x - x_{k+1}) \cdots (x - x_n).$$

In other words, the coefficients of $L_k(x)$ are up to a factor $m'(x_k)$, those of $\frac{m(x)}{x - x_k}$.

Exercises

Recall that the dual of a vector space V over a field K is the space V^* of the linear transformations $f : V \rightarrow K$. If $\{v_i\}$ is a basis of V, the basis of V^*, that is, the dual of the basis $\{v_i\}$, consists of the f_is such that $f_i(v_j) = \delta_{i,j}$. In the following exercises we consider the space of polynomials of degree at most n we have considered in this section.

18. Determine the dual basis of that consisting of the polynomials $L_k(x)$. [*Hint*: $f_i : L_k(x) \rightarrow L_k(x_i)$.]

19. Determine the dual basis of the standard basis $1, x, \ldots, x^n$. [*Hint*: consider f_i mapping a polynomial to its i-th coefficient.]

20. Prove that the polynomials $(x - a)^k$, $k = 0, 1, \ldots, n$, make up a basis. Which is its dual basis?

21. Prove that every set of polynomials $p_k(x)$ of degrees $k = 0, 1, \ldots, n$ constitute a basis.

22. Prove that the Vandermonde matrix is non singular by using Corollary 1.2.

23. Prove that, besides the $2n^2 + n - 1$ multiplications, computing $u(x)$ by the Lagrange method requires n additions, $2n^2 + 2$ subtractions and $n + 1$ divisions.

1.4.2 Newton's method

The interpolating polynomial can be computed, as for integers, in a Newtonian way. If

$$L_0^{(r)}, L_1^{(r)}, \ldots, L_r^{(r)},$$

are the integers provided by Lagrange method in the case of $r + 1$ moduli $m_0, m_1, \ldots, m_r$, and of $u_0, u_1, \ldots, u_r$ are $r + 1$ arbitrary integers, a solution is given by:

$$u^{(r)} = u_0 L_0^{(r)} + u_1 L_1^{(r)} + \cdots + u_r L_r^{(r)}.$$

Consider the difference $u^{(n)} - u^{(n-1)}$; we have:

$$u^{(n)} - u^{(n-1)} = u_0(L_0^{(n)} - L_0^{(n-1)}) + \cdots + u_{n-1}(L_{n-1}^{(n)} - L_{n-1}^{(n-1)}) + u_n L_n^{(n)}.$$

Now, $L_k^{(n)}$ and $L_k^{(n-1)}$ are both congruent to 1 modulo m_k and to 0 modulo m_i, $i \neq k$, $k = 0, 1, \ldots, n-1$, so the difference $u^{(n)} - u^{(n-1)}$ is congruent to 0 modulo $m_0, m_1, \ldots, m_{n-1}$, hence it is so modulo their product q_n as well:

$$u^{(n)} - u^{(n-1)} \equiv 0 \bmod q_n,$$

or $u^{(n)} = u^{(n-1)} + a_n q_n$. Analogously, $u^{(n-1)} = u^{(n-2)} + a_{n-1} q_{n-1}$ and

$$u = u^{(n)} = a_0 + a_1 q_1 + \cdots + a_n q_n. \tag{1.7}$$

The situation is pretty analogous in the case of polynomials. We consider the Lagrange polynomials for the case of $r + 1$ points: $x_0, x_1, \ldots, x_r$, and $r + 1$ values $u(x_0), u(x_1), \ldots, u(x_r)$ of $u(x)$:

$$L_0^{(r)}(x), L_1^{(r)}(x), \ldots, L_r^{(r)}(x).$$

The solution in the case of $n + 1$ points is

$$u(x) = u^{(n)}(x) = u_0 L_0^{(n)}(x) + \cdots + u_{n-1} L_{n-1}^{(n)}(x) + u_n L_n^{(n)}(x).$$

For $k = 0, 1, \ldots, n-1$, the value of $L_k^{(n)}(x)$ and $L_k^{(n-1)}(x)$ at $x = x_k$ is 1, while at $x = x_j$, $j \neq k$, it is 0. $L_n^{(n)}$ is zero at the same points. From this follows that the difference $u^{(n)}(x) - u^{(n-1)}(x)$ is divisible by $(x - x_0)(x - x_1) \cdots (x - x_{n-1})$, so:

$$u^{(n)}(x) = u^{(n-1)}(x) + a_n(x - x_0)(x - x_1) \cdots (x - x_{n-1}).$$

This argument leads to the *Newton interpolation formula*:

$$u(x) = u^{(n)}(x) = a_0 + a_1(x - x_0) + a_2(x - x_0)(x - x_1) + \cdots \tag{1.8}$$
$$+ a_n(x - x_0)(x - x_1) \cdots (x - x_{n-1}).$$

The analogy with formula (1.3) is clear. Moreover, as in (1.3), the coefficient a_k of (1.8) does not depend on $x_{k+1}, \ldots, x_n$ nor on $u_{k+1}, \ldots, u_n$.
The computation of the a_is is done as in the case of integers:

$$u_0 = u(x_0) = a_0,$$
$$u_1 = u(x_1) = a_0 + a_1(x_1 - x_0),$$

and hence:

$$a_1 = \frac{u_1 - u_0}{x_1 - x_0}.$$

Moreover,

$$u_2 = u(x_2) = a_0 + a_1(x_2 - x_0) + a_2(x_2 - x_0)(x_2 - x_1)$$
$$= u_0 + \frac{u_1 - u_0}{x_1 - x_0}(x_2 - x_0) + a_2(x_2 - x_0)(x_2 - x_1),$$

so:

$$a_2 = \frac{(u_2 - u_0)(x_1 - x_0) - (u_1 - u_0)(x_2 - x_0)}{(x_1 - x_0)(x_2 - x_0)(x_2 - x_1)},$$

and so on.

We have mentioned an analogy among (1.3) and (1.8), but they are actually the same formula. Taking an integer mod m_k means taking the remainder of its division by m_k. Taking a polynomial mod $(x - x_k)$ means taking the remainder of its division by $x - x_k$, so its value in x_k. In particular, the remainder of the division of $x - x_i$ by $x - x_k$ is $x_k - x_i$. From this viewpoint the two formulas are actually the same. Indeed, in the case of (1.3), $a_0 \equiv u_0$ mod m_0, that is, a_0 is the remainder of the division of u_0 by m_0. For a_1, (1.3) gives:

$$a_1 \equiv \frac{u_1 - u_0}{m_0} \text{ mod } m_1, \tag{1.9}$$

while (1.8) gives:

$$a_1 = \frac{u_1 - u_0}{x_1 - x_0}. \tag{1.10}$$

Now, $x_1 - x_0$ is the remainder of the division of $x - x_0$ by $x - x_1$, which corresponds to m_0 mod m_1 in (1.9). The situation is even clearer if we write (1.9) in the form:

$$a_1 \equiv \frac{(u_1 - u_0) \text{ mod } m_1}{m_0 \text{ mod } m_1}, \tag{1.11}$$

and (1.10) in the form:

$$a_1 \equiv \frac{(u_1 - u_0) \text{ mod } (x - x_1)}{(x - x_0) \text{ mod } (x - x_1)}, \tag{1.12}$$

$((u_1 - u_0) \text{ mod } (x - x_1)$ is simply $u_1 - u_0$ since $u_1 - u_0$ is a constant). Analogous considerations hold for the remaining a_is.

Example. We want to determine the polynomial $u(x)$ of degree at most 3 such that, for

$$x_0 = -2, \ x_1 = -1, \ x_2 = 0, \ x_3 = 1,$$

the following holds:

$$u_0 = u(-2) = -1, \ u_1 = u(-1) = 2, \ u_2 = u(0) = 1, \ u_3 = u(1) = -2.$$

The numbers a_i appearing in (1.8) are in this case:

$$a_0 = u_0 = -1; \ a_1 = \frac{u_1 - u_0}{x_1 - x_0} = \frac{3}{1} = 3; \ a_2 = \frac{2 \cdot 1 - 3 \cdot 2}{1 \cdot 2 \cdot 1} = \frac{-4}{2} = -2.$$

For a_3 we have:

$$u_3 = u(1) = a_0 + a_1 \cdot 3 + a_2 \cdot 6 + a_3 \cdot 6 = -1 + 3 \cdot 3 - 2 \cdot 6 + a_3 \cdot 6,$$

which yields $a_3 = \frac{1}{3}$. From this it follows:

$$u(x) = -1 + 3(x + 2) - 2(x^2 + 3x + 2) + \frac{1}{3}(x^3 + 3x^2 + 2x)$$

$$= \frac{1}{3}x^3 - x^2 - \frac{7}{3}x + 1.$$

1.4.3 Divided differences

The a_is in Newton's method are a particular case of *divided differences*, in the sense we are about to discuss. Let $f(x)$ be an arbitrary function (not necessarily a polynomial), and let $y_0, y_1, \ldots, y_n$ be its values at the points $x_0, x_1, \ldots, x_n$. The fraction $\frac{y_i - y_j}{x_i - x_j}$ is denoted by $[x_i x_j]$:

$$[x_0 x_1] = \frac{y_0 - y_1}{x_0 - x_1}, \quad [x_1 x_2] = \frac{y_1 - y_2}{x_1 - x_2}, \ldots.$$

Note that the value of $[x_i x_j]$ does not depend on the ordering of its arguments: $[x_i x_j] = [x_j x_i]$. The numbers $[x_i x_j]$ are called *first-order divided differences* of the function $f(x)$. The fractions:

$$[x_0 x_1 x_2] = \frac{[x_0 x_1] - [x_1 x_2]}{x_0 - x_2},$$

and

$$[x_1 x_2 x_3] = \frac{[x_1 x_2] - [x_2 x_3]}{x_1 - x_3},$$

and so on are the *second-order divided differences*. In general,

$$[x_0 x_1 \ldots x_n] = \frac{[x_0 x_1 \ldots x_{n-1}] - [x_1 x_2 \ldots x_n]}{x_0 - x_n}$$

are the divided differences of order n. By definition, $[x_i] = y_i$, the value $f(x)$ at x_i; these are the *divided differences of order* 0. As in the case $n = 1$, the differences of order n do not depend on the ordering of the x_is: $[x_0 x_1 \ldots x_k] = [x_{i_0} x_{i_1} \ldots x_{i_k}]$, where $i_0, i_1, \ldots, i_k$ is an arbitrary permutation of $0, 1, \ldots, k$.

The following table shows how to compute the first few differences ($n = 4$):

$$
\begin{array}{ll}
x & y \\[4pt]
x_0 \ y_0 & \\
 & [x_0 x_1] \\
x_1 \ y_1 & \qquad [x_0 x_1 x_2] \\
 & [x_1 x_2] \qquad\qquad [x_0 x_1 x_2 x_3] \\
x_2 \ y_2 & \qquad [x_1 x_2 x_3] \qquad\qquad\qquad [x_0 x_1 x_2 x_3 x_4] \\
 & [x_2 x_3] \qquad\qquad [x_1 x_2 x_3 x_4] \\
x_3 \ y_3 & \qquad [x_2 x_3 x_4] \\
 & [x_3 x_4] \\
x_4 \ y_4 &
\end{array}
$$

The value of a term in square brackets [...] can be obtained by dividing the difference between the two bracketed terms on its left by the difference between the first and last x in the brackets. In other words: the value of a term in square brackets is obtained by taking the difference between the two bracketed terms obtained by deleting two different values of x and dividing it by the difference of those values.

Example.

$$
\begin{array}{ll}
x & y \\[4pt]
-2 \ -1 & \\
 & \quad 3 \\
-1 \quad 2 & \qquad -2 \\
 & \quad -1 \qquad \frac{1}{3} \\
0 \quad 1 & \qquad -1 \\
 & \quad -3 \\
1 \ -2 &
\end{array}
$$

In the case when $f(x)$ is a polynomial $u(x)$, as in the previous section, it is immediate to check that the terms a_i of (1.8) are given by:

$$a_0 = y_0, a_1 = [x_0 x_1], \ldots, a_i = [x_0 x_1 \ldots x_i], \ldots, a_n = [x_0 x_1 \ldots x_n].$$

So, computing a_i in this way requires n divisions for the differences of order 1, $n-1$ for those of order 2, ..., 1 division for that of order n; all in all, $\frac{1}{2}(n^2 + n)$ divisions, plus $n^2 + n$ subtractions, saving about 3/4 of the operations required by the Lagrange method.

1.5 Applications

Let us see now an application of the Chinese remainder theorem to a problem in probability theory, and three examples of applications of Lagrange polynomials: one to a problem about matrices, one to the computation of the product of two polynomials, and a third one to a problem in cryptography.

1. It can be proved that it is possible to define a measure on the integers Z, that is, a function μ defined on the subsets S of Z with values in $[0,1]$ and such that:

i) $\mu(Z) = 1$;

ii) $\mu(i+S) = \mu(S)$, $i \in Z$, (invariance by translations);

iii) $\mu(S \cup T) = \mu(S) + \mu(T)$, if $S \cap T = \emptyset$.

(Property iii) – finite additivity – extends to disjoint finite unions.) These three properties make it possible to define a probability on Z: for all $S \subseteq Z$, the probability that an element x of Z belongs to S is given by the measure of S:

$$\mu(S) = p(x \in S).$$

Let m be a positive integer, and let $S_0, S_1, \ldots, S_{m-1}$ be the congruence classes mod m. Since $S_i = i + S_0$, from ii) we get $\mu(S_i) = \mu(S_0)$, so all congruence classes mod m have the same measure, and hence, keeping in mind i) and iii),

$$1 = \mu(Z) = \mu(S_0 \cup S_1 \cup \ldots \cup S_{m-1}) = m\mu(S_i).$$

So all S_is have measure $\mu(S_i) = 1/m$. In other words, this means that, given a and m in Z, the probability that an integer x is congruent to a mod m (that is, that x belongs to the same class S_i as a), is $1/m$:

$$p(x \equiv a \bmod m) = \frac{1}{m}.$$

Let now $m_0, m_1, \ldots, m_n$ be pairwise coprime integers. The Chinese remainder theorem tells us that every congruence system:

$$x \equiv a_i \bmod m_i, \; i = 0, 1, \ldots, n, \tag{1.13}$$

is equivalent to a single congruence:

$$x \equiv a \bmod m, \tag{1.14}$$

where m is the product of the m_is. By the above, the probability that (1.14) holds is $1/m$, that is, $\frac{1}{m_0} \cdot \frac{1}{m_1} \cdots \frac{1}{m_n}$. In other words, the probability that (1.13) simultaneously hold (that is, the probability that (1.14) holds) is the product of the probabilities of each of them holding; we may conclude that *congruences whose moduli are pairwise coprime are statistically independent.*

2. If $f(x) = a_0 x^n + a_1 x^{n-1} + \cdots + a_{n-1}x + a_n$ is a polynomial and A is a square matrix, we mean by $f(A)$ the matrix obtained by summing up the matrices $a_k A^k$ (whose entries are those of the k-th power of A multiplied by a_k), for $k = 0, 1, \ldots, n$ (the matrix A^0 is the identity matrix I). It is possible to prove that, given a matrix A, there is always a polynomial $f(x)$ such that $f(A) = 0$, where 0 is the zero matrix, the matrix all of whose entries are zero. (For instance, by the Hamilton-Cayley theorem, one such polynomial is the characteristic polynomial of A, that is, $det(A - xI)$, which has degree n if A is a $n \times n$ matrix.) So, among those polynomials there is one with least degree $m + 1$ and monic, $m(x)$, the *minimal polynomial of A*. Assume that all of the roots of $m(x)$, $\lambda_0, \lambda_1, \ldots, \lambda_m$, are distinct. In this case we may consider the Lagrange polynomials $L_k(x)$, $k = 0, 1, \ldots, m$, with respect to the λ_is, and the matrices $A_k = L_k(A)$. We know that the polynomials $L_k(x)$ are orthogonal idempotents and that they sum to 1, and this can now be expressed by the following three relations:

1. $I = A_0 + A_1 + \cdots + A_m$,
2. $A_i A_j = 0$, $i \neq j$,
3. $A_k^2 = A_k$, for all k.

(Formulas 2. and 3. are derived from congruences mod $m(x)$, which become now congruences mod $m(A)$; but $m(A) = 0$, so the congruences are identities.)

The matrix A represents a linear transformation of a vector space V. Let us show how 1., 2., and 3. allow us to decompose the space into a direct sum of subspaces. If $v \in V$, we have, by 1.,

$$v = Iv = A_0 v + A_1 v + \cdots + A_m v,$$

so

$$V = A_0(V) + A_1(V) + \cdots + A_m(V),$$

and V is the sum of the subspaces $A_i(V)$. Moreover, this sum is direct. Indeed, if

$$A_i v_i = A_0 v_0 + \cdots + A_{i-1} v_{i-1} + A_{i+1} v_{i+1} \cdots + A_m v_m,$$

for some $v_0, v_1, \ldots, v_m$, then, by multiplying by A_i and keeping in mind 2. and 3., we have $A_i v_i = 0$.

Let us prove now that if $v \in A_i(V)$, then $Av = \lambda_i v$, for all i. We know that

$$x = \lambda_0 L_0(x) + \lambda_1 L_1(x) + \cdots + \lambda_m L_m(x);$$

for $x = A$ this becomes:

$$A = \lambda_0 A_0 + \lambda_1 A_1 + \cdots + \lambda_m A_m.$$

Now, if $v = A_i(u)$, we have $(A - \lambda_i I)v = (A - \lambda_i I)A_i u = A A_i u - A_i \lambda_i u$. The last quantity is equal to zero. Indeed, from the expression of A given here, we have $A A_i = \lambda_i A_i^2 = \lambda_i A_i$, and hence $A A_i u - A_i \lambda_i u = A_i \lambda_i u - A_i \lambda_i u = 0$. So $Av - \lambda_i v = (A - \lambda_i I)v = 0$, and $Av = \lambda_i v$, as required.

Remark. We have proved that $A_i(V) \subseteq Ker(A - \lambda_i I)$. But the sum of the subspaces $K_i = Ker(A - \lambda_i I)$ is direct too (since eigenvectors corresponding to distinct eigenvalues are independent), so, by a dimension argument, V is direct sum of the

K_is. Choosing a basis in each of these subspaces, the matrix A can be transformed into a diagonal block matrix: each block is scalar, since it corresponds to one of the K_is, so the matrix has entries λ_i on the main diagonal and zero elsewhere. Thus, we have that if the minimal polynomial of a matrix A can be split into distinct linear factors over the coefficient field, then A is diagonalisable. (The converse is also true.)

3. If $g(x)$ and $h(x)$ are two polynomials, having degree n, then the product $f = gh$ can be computed as follows. Let $x_0, x_1, \ldots, x_{2n}$ be distinct points; then the products $y_i = g(x_i)h(x_i)$ yield $2n + 1$ values of f, which has degree $2n$, so they determine f. The explicit expression of f can be then obtained by the Lagrange method starting from the x_is and the y_is. If g and h have different degrees, $m = \partial g < \partial h = n$, we may make g artificially "of degree n" by using monomials of the form $0 \cdot x^k$.

Computing a polynomial of degree n at a point requires a number of multiplications on the order of n, so, to find the values at n points, or at a number of points that is a linear function of n, we need a number of multiplications on the order of n^2. We shall see later (Chapter 5) that, by choosing as points the n-th roots of unity, the number of multiplications can be lowered to $n \log n$.

4. This example is about dividing a secret D (for instance, a number giving the combination to open a safe) in n parts $D_0, D_1, \ldots, D_{n-1}$ in such a way that:

i)   knowing k among the D_is it is possible to compute D;
ii) knowing $k - 1$ among the D_is leaves D undetermined, in the sense that all of its possible values are still equally likely.

To this end, consider a polynomial of degree $k - 1$:

$$p(x) = a_0 + a_1 x + \cdots + a_{k-1} x^{k-1},$$

where $a_0 = D$, while the remaining coefficients are chosen randomly. If $x_0, x_1, \ldots, x_{n-1}$, $n \geq k$, are pairwise distinct numbers, consider the numbers:

$$p(x_0), p(x_1), \ldots, p(x_{n-1}).$$

Now, knowing k pairs $D_i = (x_i, p(x_i))$ it is possible to determine $p(x)$ by interpolation, hence D as $p(0)$, while $k - 1$ pairs are not sufficient to determine the polynomial and hence D. In other words, to come back to our safe, if the n pairs $(x_i, p(x_i))$ are given to n people, one pair each, in order to be able to open the safe at least k people have to be together.

The same problem can be studied using modular arithmetic. Given an integer D, let p be a prime number greater than both D and n. Choose randomly the coefficients $a_1, a_2, \ldots, a_{k-1}$ of $p(x)$ among $0, 1, \ldots, p - 1$, as well as the n values x_i, with the only constraint $x_i \neq 0$. The numbers $p(x_i)$ are computed modulo p. Now, knowing $k - 1$ pairs $D_i = (x_i, p(x_i))$, for each possible value of D it is possible to construct one, and only one, polynomial $p'(x)$ such that $p'(0) = D$. All these p polynomials are by construction equally probable, so nothing transpires of the actual value of D.

References

For the material in this chapter, see [**A**] and [**C**]. For the Euclidean algorithm, [**Kn**] § 4.5.2. For the Chinese remainder theorem and Lagrange and Newton methods, see the paper by M. Lauer, *Computing by homomorphic images* in [**CA**], which is also the source of the example discussed in the text. For the Chinese remainder theorem and interpolation, § 4.6.4 of [**Kn**] and [**Li**]. Item 4 in Applications comes from [**Kn**], page 486.

2

p-adic series expansions

2.1 Expansions of rational numbers

In this section we shall deal with the expansion of a rational number as a series of powers of a not necessarily prime number $p \geq 2$. As we shall see, the techniques we shall use extend to rational numbers the way we write a positive integer number in base p. In that case, the expansion is finite, that is, we have a polynomial in p with coefficients at least 0 and at most $p - 1$; here the expansion is infinite, that is, we have a power series in p, with the same range for the coefficients.

In order to write an integer $x \geq 0$ in a given base p we divide x by p, to obtain a remainder c_0 and a quotient q: $x = c_0 + pq$. The remainder c_0 is the first digit of the expansion, and represents a "first approximation" to the number x, that is, the number x "up to multiples of p". To find the second digit, we divide q by p: $q = c_1 + pq_1$, so, by substituting the expression we have found for q, we get $x = c_0 + c_1 p + q_1 p^2$. The number $c_0 + c_1 p$ is the "second approximation" of x, that is, x "up to multiples of p^2"; and so on. Since we have $x < p^t$ for some t, if t is the least such exponent, the coefficients from c_t on are all zero: $x = c_0 + c_1 p + \cdots + c_{t-1} p^{t-1} + 0 \cdot p^t + \cdots$. We write $x = c_0.c_1 \ldots c_{t-1}$.

So our procedure is as follows: having found $c_0 \equiv x \bmod p$, in order to find c_1 consider q, that is, the integer:

$$\frac{x - c_0}{p}$$

and divide it by p; c_1 will be taken to be the remainder of this division:

$$c_1 \equiv \frac{x - c_0}{p} \bmod p.$$

For c_2, take the integer:

$$\frac{x - (c_0 + c_1 p)}{p^2}$$

Machì A.: Algebra for Symbolic Computation.
DOI 10.1007/978-88-470-2397-0_2, © Springer-Verlag Italia 2012

and divide it by p; c_2 is the remainder of this division:

$$c_2 \equiv \frac{x - (c_0 + c_1 p)}{p^2} \mod p,$$

and so on. In general, let $e_n = c_0 + c_1 p + \cdots + c_{n-1} p^{n-1}$ be the n-th order approximation of x; the number $x - e_n = R_n$ is the n-th order *remainder*. The new approximation will be $e_{n+1} = e_n + c_n p^n$, where c_n is the remainder of the division of $\frac{x - e_n}{p^n}$ by p; so, for $n > 0$, we have the general formula:

$$c_n \equiv \frac{x - (c_0 + c_1 p + \cdots + c_{n-1} p^{n-1})}{p^n} \mod p, \tag{2.1}$$

where the coefficients c_i are such that $0 \leq c_i < p$.

Let now $\frac{a}{b}$ be a rational number. The following theorem holds.

Theorem 2.1. *Let $\frac{a}{b}$ be a rational number, $b \geq 1$, and $p \geq 2$ an integer with $(p, b) = 1$. Then there exists a unique pair of integers c, d, with $0 \leq c < p$ and such that:*

$$\frac{a}{b} = c + \frac{d}{b} p. \tag{2.2}$$

Proof. Let $bx + py = 1$. Then $b(ax) + p(ay) = a$. If $0 \leq ax < p$, we get the claim by taking $c = ax$ and $d = ay$. Otherwise, divide ax by p; we get $ax = pq + r$, with $0 \leq r < p$, and having set:

$$c = ax - pq \quad \text{and} \quad d = ay + bq,$$

we have:

$$bc + pd = a;$$

hence, dividing by b, we get (2.2). (Note that $c \equiv ab^{-1} \mod p$, and that b^{-1} exists mod p since $(p, b) = 1$.) This proves existence. As for uniqueness, let c', d' be another pair with the required properties; then $bc' + pd' = a$, so, subtracting from the former, $b(c - c') = p(d' - d)$. From this follows that p divides $b(c - c')$, and since $(p, b) = 1$, p divides $c - c'$. But c and c' are both less than p and non-negative, so $-p < c - c' < p$. It follows $c - c' = 0$ and $c = c'$, and hence also $d = d'$. $\diamond$

Note that in the case of an integer value of $\frac{a}{b}$, that is, when $b = 1$, formula (2.2) becomes $a = c + dp$, so c is the remainder and d the quotient of the division of a by p.

The integer c in (2.2) is the first approximation of a/b. Set $c = c_0$, and apply the theorem to the pair d, b. We find, as above,

$$\frac{d}{b} = c_1 + \frac{d_1}{b} p,$$

and analogously:

$$\frac{d_1}{b} = c_2 + \frac{d_2}{b}p.$$

Going on like this, and substituting at each step $\frac{d_i}{b}$ with its expression $c_{i+1} + \frac{d_{i+1}}{b}p$, we find the expansion:

$$\frac{a}{b} = c_0 + c_1 p + c_2 p^2 + \cdots + c_{n-1}p^{n-1} + \frac{d_{n-1}}{b}p^n, \tag{2.3}$$

for $n = 1, 2, \ldots$ From this follows:

$$a = b(c_0 + c_1 p + \cdots + c_{n-1}p^{n-1}) + d_{n-1}p^n, \tag{2.4}$$

and hence:

$$a \equiv b(c_0 + c_1 p + \cdots + c_{n-1}p^{n-1}) \bmod p^n, \tag{2.5}$$

where $0 \le c_i < p$. So, for all $n = 1, 2, \ldots$, we have one of the congruences (2.5). It is usual to write, symbolically:

$$\frac{a}{b} = c_0 + c_1 p + c_2 p^2 + \cdots = \sum_{k \ge 0} c_k p^k,$$

or:

$$\frac{a}{b} = c_0.c_1 c_2 \ldots.$$

However, let us emphasise the fact that this expressions are just a symbolic way to write down the infinite sequence of congruences given by (2.5). The term $R_n = d_{n-1}p^n$ in (2.4) is the n-th order remainder of the expansion.

The expansion of a rational number as a power series in p is so reduced to the solution of congruences (2.5) for $n = 1, 2 \ldots$.

Let us sum up the above in the following:

Theorem 2.2. *Every rational number $\frac{a}{b}$, with $b \ge 1$, can be expanded as a series of positive powers of an integer number $p \ge 2$ such that $(p, b) = 1$, and this can be done with arbitrary precision, in the sense that the remainder R_n in (2.4) can be, by suitably carrying on the expansion, made divisible by an arbitrarily high power of p.*

The expansion given in Theorem 2.2 is called the *p-adic expansion* of the rational number a/b.

If we denote by $e_n = c_0 + c_1 p + \cdots + c_{n-1}p^{n-1}$ the n-th order approximation of a/b given by (2.5), we get

$$\frac{a}{b} \equiv e_n \bmod p^n,$$

or $be_n - a \equiv 0 \bmod p^n$. In other words, e_n is a solution of the equation $bx - a = 0$ in the integers mod p^n. So we have an equivalent way of stating Theorem 2.1.

Theorem 2.3. *If a, b and p are as in Theorem 2.1, then the equation* $bx - a = 0$ *admits a solution in integers modulo* p^n, *for all n.*

Example. Determine the 7-adic expansion of 1/12. We have:

$$12 \cdot 3 + 7 \cdot -5 = 1, \tag{2.6}$$

and hence:

$$\frac{1}{12} = 3 - \frac{5}{12}7.$$

From (2.6) multiplied by -5 we get $12 \cdot -15 + 7 \cdot 25 = -5$. Since -15 is not included between 0 and 7, divide it by 7: $-15 = 7 \cdot -3 + 6$. From this follows $12 \cdot 6 + 7 \cdot -11 = -5$, or

$$-\frac{5}{12} = 6 - \frac{11}{12}7.$$

Analogously, $12 \cdot -33 + 7 \cdot 55 = -11$, $-33 = 7 \cdot -5 + 2$ and $12 \cdot 2 + 7 \cdot (55 + 12 \cdot -5) = 12 \cdot 2 + 7 \cdot -5 = -11$, and

$$-\frac{11}{12} = 2 - \frac{5}{12}7.$$

We restart with $-5/12$, so the following digit will be 6 again. From this follows:

$$\frac{1}{12} = 3.6262 \ldots \text{ mod } 7.$$

So the expansion is *periodic*, with period 2 (the last two digit 6 and 2). We write this as:

$$\frac{1}{12} = 3.\overline{62} \text{ mod } 7.$$

In the words of Theorem 2.3 we have then that the equation $12x - 1 = 0$ has solution 3 in integers mod 7, solution $3 + 6 \cdot 7 = 45$ in integers mod 7^2 (indeed, $12 \cdot 45 - 1 = 540 - 1 = 539 = 11 \cdot 7^2 \equiv 0 \text{ mod } 7^2$), and so on.

So the procedure is pretty analogous to the one for integers. More precisely, from $\frac{a}{b} = c_0 + \frac{d}{b}p$ we have:

$$\frac{\frac{a}{b} - c_0}{p} = \frac{1}{b}\frac{a - bc_0}{p} = \frac{d}{b},$$

hence $d = \frac{a - bc_0}{p}$. Now, $d = bc_1 + pd_1 \equiv bc_1 \text{ mod } p$, so:

$$c_1 \equiv db^{-1} = \frac{1}{b}\frac{a - bc_0}{p} \text{ mod } p.$$

Analogously, $d_1 = \frac{a - b(c_0 + c_1 p)}{p^2}$ and

$$c_2 = \frac{1}{b}\frac{a - b(c_0 + c_1 p)}{p^2} \text{ mod } p,$$

and in general:

$$c_n \equiv \frac{1}{b} \frac{a - b(c_0 + c_1 p + \cdots + c_{n-1} p^{n-1})}{p^n} \bmod p,$$

which coincides with (2.1) when $b = 1$. Note that the expression

$$\frac{a - b(c_0 + c_1 p + \cdots + c_{n-1} p^{n-1})}{p^n}$$

is always an integer. Indeed, it is the integer d_{n-1} of (2.3) (or (2.4)).

Denoting here too by e_n the approximation of a/b of order n:

$$e_n = c_0 + c_1 p + \cdots + c_{n-1} p^{n-1}$$

we have the following algorithm that provides the coefficients $c_0, c_1, \ldots, c_{n-1}$ or the approximate values $e_1, e_2, \ldots, e_n$, for all n:

input: a, b, p, n;
output: $c_0, c_1, \ldots c_{n-1}$ (or: $e_1, e_2, \ldots, e_n$);
$d := 1/b \bmod p$;
$c_0 := ad \bmod p$;
$e_1 := c_0$;
for k from 1 to $n - 1$ do:
$q_k := \mathrm{quotient}(a - be_k, p^k)$;
$c_k := dq_k \bmod p$;
$e_{k+1} := e_k + c_k p^k$.

The expansion of a non-negative integer number in base p is periodic, with period 1: the period consists of a single digit, zero. We shall see in a moment that a negative integer number has a periodic expansion of period 1 too, but the repeating digit is $p - 1$, and that in general the series expansion of a rational number is always periodic, that is, of the form:

$$\frac{a}{b} = c_0.c_1 \ldots c_k \overline{c_{k+1} c_{k+2} \ldots c_{k+d}},$$

where this expression means that $c_{k+1} = c_{k+d+1} = c_{k+2d+1} = \ldots$, and analogously for the other c_is. If d is the least length for which we have these inequalities, then we say that a/b is periodic of period d. Conversely, a periodic expansion represents a rational number (see Theorem 2.5 below).

Consider now an expansion of the form:

$$\frac{a}{b} = a_0 + a_1 p + a_2 p^2 + \cdots,$$

where a_i are arbitrary integers, and let us see how to convert this expansion into the form in which the coefficients are between 0 and $p - 1$, a form we shall call *reduced*. If a_0 is not a number between 0 and $p - 1$, divide it by p, $a_0 = pq_1 + r_1$, and substitute in the expansion $a_0 - pq_1$ for a_0. Having subtracted pq_1 we have to add it again, and we obtain:

$$(a_0 - pq_1) + (a_1 + q_1)p + a_2 p^2 + \cdots,$$

and continuing, $a_1 + q_1 = pq_2 + r_2$ and

$$(a_0 - pq_1) + (a_1 + q_1 - pq_2)p + (a_2 + q_2)p^2 + \cdots.$$

So,

$$\frac{a}{b} = c_0 + c_1 + \cdots,$$

where $c_i = a_i + q_i - pq_{i+1}$, and $0 \le c_i < p$ since those are remainders in divisions by p.

If an expansion has the form $a_0.a_1 a_2 \ldots$, it can be rewritten in reduced form by substituting each digit, starting from the first one, with the remainder we get dividing it by p and adding the quotient we obtain (the "carry") to the next digit. This applies in particular to negative a_is.

Examples. 1. Let $7.530860\overline{0}$ an integer mod 3. It can be successively transformed as follows:

$$7.530860\overline{0} = 1.730860\overline{0} = 1.150860\overline{0}$$
$$= 1.121860\overline{0} = 1.121280\overline{0}$$
$$= 1.121222\overline{0},$$

which is the base-3 expression of 2155: $1 + 1 \cdot 3 + 2 \cdot 3^2 + \cdots + 2 \cdot 3^6 + 0 \cdot 3^7 + 0 \cdot 3^8 + \cdots$

2. Consider, again in base 3, the negative number -2155. By the previous example, we have $-2155 = -1 - 1 \cdot 3 - 2 \cdot 3^2 - \cdots - 2 \cdot 3^6 + 0 \cdot 3^7 + 0 \cdot 3^8 + \cdots$. Here the coefficients are less than 3 in absolute value, so dividing by p always yields a quotient -1: $-k = p \cdot -1 + (p - k)$. Thus, the reduction is performed by substituting $a_k + p, a_{k+1} - 1$ for two consecutive coefficients a_k, a_{k+1}. So we have: $-2155 = 2 - 2 \cdot 3 - 2 \cdot 3^2 - \cdots$, and going on:

$$-2155 = 2 + 1 \cdot 3 + 0 \cdot 3^2 - 2 \cdot 3^3 + \cdots.$$

When we arrive at $2 + 1 \cdot 3 + \cdots + 1 \cdot 3^5 - 3 \cdot 3^6 + 0 \cdot 3^7 + \cdots$, we have: $\cdots - 3 \cdot 3^6 + 0 \cdot 3^7 + 0 \cdot 3^8 + \cdots = \cdots 0 \cdot 3^6 - 1 \cdot 3^7 + 0 \cdot 3^8 + \cdots = \cdots 2 \cdot 3^7 - 1 \cdot 3^8 + \cdots$. In other words, the coefficients of the powers from 3^7 on are all equal to 2. Note that the coefficient of 3^6 is the last coefficient of 2155 before the repeating 0. We may conclude that -2155 is periodic, with period 1. Hence, write $-2155 = 2.101000\overline{2}$. The repeating digit is 2, that is, $3 - 1$. This fact is general:

Theorem 2.4. *A series expansion modulo p in which the coefficients from some point on are all equal to $p - 1$ represents a negative integer number, and conversely.*

Proof. Let $x = c_0.c_1c_2\ldots c_k\ p-1\ p-1\ldots$ be such an expansion. We may suppose that it is a reduced expansion, since a reduction procedure, if any, just moves rightwards the beginning of the repeating part. So, $-x = -c_0 - c_1p - \cdots - c_kp^k + (1-p)p^{k+1} + (1-p)p^{k+2} + \cdots$. Reducing, we get $\cdots + (p-c_k-1)p^k + (1-p-1)p^{k+1} + \cdots = \cdots + 0\cdot p^{k+1} + (1-p-1)p^{k+2} + \cdots$ and so on, so $c_i = 0$ for $i > k$. From this follows that $-x$ is a positive integer, and hence x is a negative integer. Conversely, if x is negative, $-x$ is positive, so $-x = d_0, d_1d_2\ldots\bar{0}$, and hence $x = -d_0 - d_1p - \cdots - d_kp^k + 0\cdot p^{k+1} + \cdots$. In the usual reduction we have: $\cdots + (p-d_k-1)p^k - 1\cdot p^{k+1} + 0\cdot p^{k+2} + \cdots = \cdots (p-1)p^{k+1} - 1\cdot p^{k+2} + \cdots$. So we notice that all the coefficients of the powers p^i with $i > k$ are equal to $p-1$. ◇

Examples. **1.** Determine the p-adic expansion of

$$\frac{1}{1-p}.$$

With the method of Theorem 2.1 we have:

$$1\cdot(1-p) + 1\cdot p = 1,$$

and hence:

$$\frac{1}{1-p} = 1 + \frac{1}{1-p}p,$$

and $c_0 = 1$. From this follows

$$\frac{1}{1-p} = 1 + (1 + \frac{1}{1-p}p)p = 1 + p + \frac{1}{1-p}p^2,$$

and $c_1 = 1$. Going on like this, or by induction, we find:

$$\frac{1}{1-p} = 1 + p + p^2 + \cdots.$$

2. More generally, let us expand

$$\frac{1}{1-p^n},$$

this time by running the algorithm. We have:

$$1 \equiv (1-p^n)c_0 = c_0 - p^nc_0 \bmod p,$$

and hence $c_0 = 1$. Moreover,

$$1 \equiv (1-p^n)(1+c_1p) = 1 - p^n + c_1p - c_1p^{n-1} \bmod p^2.$$

If $n \geq 2$, we have $c_1 \equiv 0 \bmod p$. Analogously, all c_is with $i < n$ are equal to zero. So:

$$1 \equiv (1-p^n)(1+c_np^n) = 1 + c_np^n - p^n - c_np^{2n} \bmod p^{n+1},$$

and hence $0 \equiv c_n - 1 \bmod p$, and $c_n = 1$. As above, all coefficients from c_{n+1} to c_{2n-1} are zero, and $c_{2k} = 1$. In conclusion, $c_i = 1$ for i a multiple of n, and $c_i = 0$ otherwise:

$$\frac{1}{1-p^n} = 1 + p^n + p^{2n} + \cdots + p^{kn} + \cdots .$$

Remark. The expansions in these examples can be found as particular cases of the formal expansion:

$$\frac{1}{1-x} = 1 + x + x^2 + \cdots ,$$

by which we mean the sequence of congruences:

$$1 \equiv (1-x)(1 + x + \cdots + x^{n-1}) \bmod x^n$$

(indeed, the left-hand side equals $1 - x^n \bmod x^n$, and this is equivalent to $1 \bmod x^n$, $n = 1, 2, \ldots$; by "mod x^n" we mean "when we set $x^n = 0$"). With $x = p$, or $x = p^n$, we obtain the above expressions, as well as, with $x = a$, $a \neq 1$, every expression $\frac{1}{1-a^n}$. We shall come back to this fact when we shall discuss series expansions of rational functions.

Let us see now a standard theorem.

Theorem 2.5. *A p-adic expansion represents a rational number if and only if it is periodic.*

Proof. Assume first we have a *pure* periodic number (that is, the repeating part starts with the first digit) of period d:

$$x = \overline{c_0.c_1 c_2 \ldots c_{d-1}} ,$$

or:

$$x = c_0 + c_1 p + \cdots + c_{d-1}p^{d-1} + c_0 p^d + \cdots + c_{d-1}p^{2d-1} + \cdots .$$

Factoring p^d out, this expression can be written as:

$$x = c_0 + c_1 p + \cdots + c_{d-1}p^{d-1} + (c_0 + c_1 p + \cdots + c_{d-1}p^{d-1})p^d + \cdots ,$$

that is:

$$x = (c_0 + c_1 p + \cdots + c_{d-1}p^{d-1})(1 + p^d + p^{2d} + \cdots),$$

and keeping in mind that the right-hand series sums to $1/(1-p^d)$,

$$x = c_0 + c_1 p + \cdots + c_{d-1}p^{d-1} \cdot \frac{1}{1-p^d} = \frac{c_0 + c_1 p + \cdots + c_{d-1}p^{d-1}}{1-p^d} .$$

So the number x is a ratio of two integers, so it is rational. Note that in this way we get a negative rational number (since $1 - p^d$ is negative), which is a proper fraction, as $1 - p^d$ is greater (in absolute value) than the numerator (by Exercise 7, we have an equality if and only if all c_is are equal to $p - 1$, so $x = -1$, an integer), while the denominator is coprime with p. Moreover, d is the smallest exponent such that $x(1 - p^d)$ is periodic, otherwise the period would be less than d.

Conversely, let $-\frac{a}{b}$ be a negative rational number (reduced to lowest terms) with $(b, p) = 1$, and let d be the order of p modulo b, that is, the least positive integer such that $p^d \equiv 1 \bmod b$.[1] So $p^d - 1 = bt$, for some positive integer t, $b = \frac{p^d - 1}{t}$, and hence:

$$-\frac{a}{b} = -\frac{at}{p^d - 1} = \frac{m}{1 - p^d},$$

where we have set $m = at$, which is an integer. Now, $m = at < bt$ since $a < b$, so $m < p^d - 1$; m has a base p expression that reaches p^{d-1}: $m = c_0 + c_1 p + \ldots + c_{d-1} p^{d-1}$, $0 \le c_i < p$. From this follows

$$-\frac{a}{b} = \frac{m}{1 - p^d} = m(1 + p^d + p^{2d} + \cdots)$$

$$= (c_0 + c_1 p + \cdots + c_{d-1} p^{d-1})(1 + p^d + p^{2d} + \cdots)$$

$$= \overline{c_0 . c_1 c_2 \ldots c_{d-1}},$$

so our rational number is a purely periodic number.

For the general case, let $x = a_0 . a_1 \cdots a_k \overline{c_0 c_1 \ldots c_{d-1}}$, and let s be the positive integer $s = a_0 . a_1 \cdots a_k$. By subtracting s from x, we get $x - s = 0.00 \ldots 0 \overline{c_0 c_1 \ldots c_{d-1}}$, and from it, multiplying by $p^{-(k+1)}$, we get $p^{-(k+1)}(x - s) = \overline{c_0 . c_1 c_2 \ldots c_{d-1}}$. By the above, the right-hand side is a rational number, so the left-hand one is too. From this follows that x is rational.

Conversely, if x is a rational number whose denominator is coprime with p, multiplying x by a suitable power of p and adding or subtracting an integer, we obtain a negative rational number. $\diamond$

Remarks. 1. It follows that the period of a rational number a/b is 1 if and only if the order of $p \bmod b$ is 1, that is, if and only if $p \equiv 1 \bmod b$. If a/b is an integer, that is if $b = 1$, then the order of $p \bmod 1$ is 1. Indeed, since two arbitrary integers are always congruent mod 1, the least positive integer d such that $p^d \equiv 1 \bmod b$ is 1. So we have found again the fact that the period of an integer is 1.

2. The negative rational numbers $\frac{a}{b}$, $a < b$, are always represented by purely periodic numbers. For the positive ones, there are always further digits before the period.

Examples. 1. Let us expand $-\frac{1}{3} \bmod 5$. We have $5^2 \equiv 1 \bmod 3$, so the period is $d = 2$. Moreover, $5^2 - 1 = 24 = 3 \cdot 8$, so $t = 8$, $m = 1 \cdot 8 = 8$, $8 = 3 + 1 \cdot 5$, and:

$$-\frac{1}{3} = \frac{3 + 1 \cdot 5}{1 - 5^2} = (3 + 1 \cdot 5)(1 + 5^2 + 5^4 \cdots)$$

$$= 3 + 1 \cdot 5 + 3 \cdot 5^2 + 1 \cdot 5^3 + \cdots$$

$$= \overline{3.1}.$$

---

[1] Since p and b are relatively prime, $p \bmod b$ is in the group of the invertible elements of Z/bZ, and as such has an order, that is, the least integer d such that the element raised to the dth power equals unity.

Let us expand now $\frac{1}{3}$ mod 5. By writing $\frac{1}{3} = 2 - \frac{1}{3}5$ and using the previous example, we have $5 \cdot -\frac{1}{3} = 3 \cdot 5 + 1 \cdot 5^2 + \cdots = 0.\overline{31}$, and adding 2 we find $\frac{1}{3} = 2.\overline{31}$. Summing up (keeping into account the carries, and starting from the left), we find $2.\overline{31} + \overline{3.1} = 0.00\ldots = 0$.

2. It is quite helpful to see how to expand $\frac{1}{3}$ as a series of powers of $\frac{1}{5}$. Keeping in mind what we have seen above, dividing both the numerator and the denominator by 5^2,

$$\frac{1}{3} = \frac{3 + 1 \cdot 5}{5^2 - 1} = \frac{\frac{3}{5^2} + \frac{1}{5}}{1 - (\frac{1}{5})^2} = (\frac{1}{5} + \frac{3}{5^2})(1 + \frac{1}{5^2} + \frac{1}{5^4} + \cdots),$$

and hence:

$$\frac{1}{3} = \frac{1}{5} + \frac{3}{5^2} + \frac{1}{5^3} + \frac{3}{5^4} + \cdots = 0.\overline{13}.$$

So we see that the period is inverted with respect to $-\frac{1}{3}$, and starts after the decimal point, that is, with the positive powers of $\frac{1}{5}$. Moreover, before the point we have 0. It is easy to see that this is a general fact.

3. Let us see now an example with $p = 10$. Expand $-\frac{3}{11}$. We have $10^2 \equiv 1 \bmod 11$, so the period is $d = 2$. Moreover, $10^2 - 1 = 11 \cdot 9$, so $m = at = 3 \cdot 9 = 27$. Thus:

$$-\frac{3}{11} = \frac{-3 \cdot 9}{10^2 - 1} = \frac{27}{1 - 10^2} = \frac{7 + 2 \cdot 10}{1 - 10^2}$$
$$= (7 + 2 \cdot 10)(1 + 10^2 + 10^4 + \cdots)$$
$$= 7 + 2 \cdot 10 + 7 \cdot 10^2 + 2 \cdot 10^3 + \cdots$$
$$= \overline{7.2}.$$

Remark. The usual decimal expression of a proper fraction $\frac{a}{b}$ with a denominator coprime with 10 can be obtained as above with $p = 10$. More precisely, having found the least d such that $10^d \equiv 1 \bmod b$, that is, the period, let $10^d - 1 = bt$. Then, as above,

$$\frac{a}{b} = \frac{at}{10^d - 1} = \frac{m}{10^d - 1}.$$

Divide both the numerator and the denominator by 10^d:

$$\frac{a}{b} = \frac{\frac{m}{10^d}}{1 - (\frac{1}{10})^d}.$$

If we write the integer m in base 10, $m = c_0 + c_1 10 + c_2 10^2 + \cdots + c_{d-1} 10^{d-1}$, and expand $1 - (\frac{1}{10})^d$, we find:

$$\frac{a}{b} = \frac{c_{d-1}}{10} + \frac{c_{d-2}}{10^2} + \cdots + \frac{c_0}{10^d}(1 + \frac{1}{10^d} + \frac{1}{10^{2d}} + \cdots),$$

and hence:

$$\frac{a}{b} = 0.\overline{c_{d-1}c_{d-2}\ldots c_0}.$$

In particular, we rediscover the well-known fact that if the denominator of a fraction is coprime with 10 (that is, is not divisible by 2 nor by 5), there are no decimal digits before the repeating ones, and conversely.

Example. Let us find the usual decimal expansion of $\frac{3}{11}$. From $10^d \equiv 1 \bmod 11$, we get $d = 2$. Moreover, $10^2 - 1 = 11 \cdot 9$, so $m = at = 3 \cdot 9 = 27 = 7 + 2 \cdot 10$. From this follows:

$$\frac{3}{11} = \frac{7 + 2 \cdot 10}{10^2 - 1} = \frac{\frac{7 + 2 \cdot 10}{10^2}}{1 - \frac{1}{10^2}},$$

where the numerator and the denominator have been divided by 10^2. So:

$$\begin{aligned}
\frac{3}{11} &= \frac{7 + 2 \cdot 10}{10^2} \cdot \frac{1}{1 - \frac{1}{10^2}} \\
&= \frac{7 + 2 \cdot 10}{10^2} \cdot (1 + \frac{1}{10^2} + \frac{1}{10^4} + \cdots) \\
&= (\frac{2}{10} + \frac{7}{10^2})(1 + \frac{1}{10^2} + \frac{1}{10^4} + \cdots) \\
&= \frac{2}{10} + \frac{7}{10^2} + \frac{2}{10^3} + \frac{7}{10^4} + \cdots \\
&= 0.\overline{27}.
\end{aligned}$$

To obtain a rational number from its decimal expression, it is sufficient to follow backwards the procedure we have seen, that is, to write the period as an integer number (in our example, 27), and divide it by $10^d - 1 = 99 \ldots 9$ (that is, as many 9s as the repeating digits; in our example, 99), and finally reduce to lowest terms (in the example, $\frac{27}{99} = \frac{3}{11}$). (The expansion in powers of 10 of $-\frac{3}{11}$ we have seen above yielded $-\frac{3}{11} = \overline{7.2}$.)

Exercises

1. Which number is represented by the expansion:

$$9.999\ldots = 9.\overline{9}$$

in base 10? More in general, which number is represented by the expansion:

$$p - 1.p - 1\ p - 1 \ldots = p - 1.\overline{p - 1}$$

in base p?

2. Which is the base 10 expansion of -327?

3. Prove that the reduced form mod 3 of $1.234\ldots$ (all positive integers) is $\overline{1.202} = -\frac{61}{80}$.

4. Which number is represented by the p-adic expansion $p.\overline{p - 1}$?

5. Which number is represented by the p-adic expansion

$$p.2\,(p-1)\,3\,(p-2)\,\ldots(k+1)\,(p-k)\ldots?$$

6. Which is the necessary and sufficient condition for a p-adic expansion to represent zero?

7. Prove that $c_0 + c_1 p + \cdots + c_{n-1}p^{n-1} = p^n - 1$ if and only if all c_is are equal to $p - 1$.

8. What does it mean for a coefficient in a p-adic expansion to be zero?

9. Compute the following sum mod 5: $2.310211\overline{40} + 3.141202131\overline{0}$.

2.2 Expansions of algebraic numbers

An algebraic number is a root of a polynomial with rational coefficients. The p-adic expansion of a rational number $\frac{a}{b}$, $(p, b) = 1$, yields a method to find a root of the polynomial $p(x) = bx - a$, that is, a solution of the linear equation $bx - a \equiv 0 \bmod p^n$, for all $n = 1, 2, \ldots$ (Theorem 2.3). Let us see now the case of a quadratic equation, $x^2 - a = 0$, with p prime and p not dividing a (that is, a is not zero modulo p). Let us consider now only the case $p \neq 2$. Note first that, unlike the linear case, this equation might not have solutions even for $n = 1$. It is, for instance, the case of $x^2 - 2 = 0$ with $p = 3$. But if there is a solution mod p, then there is one mod p^n for every n. Denote this solution by $\sqrt{a}$.

Theorem 2.6. *Let p be prime and $p > 2$. If the equation $x^2 - a = 0$, $p \nmid a$, has a solution in integers modulo p, then it has one in integers modulo p^n for all n.*

Proof. As in the linear case, we have to find an expansion as a series of powers of p:

$$\sqrt{a} = c_0 + c_1 p + c_2 p^2 + \cdots .$$

Let e_1 be a solution mod p. Set $e_1 = c_0$, and suppose by induction we have found a solution e_n mod p^n in the form:

$$e_n = c_0 + c_1 p + \cdots + c_{n-1}p^{n-1},$$

that is, we have:

$$e_n^2 \equiv a \bmod p^n.$$

Thus, let us look for a solution mod p^{n+1} in the form $e_{n+1} = e_n + cp^n$. Since

$$e_{n+1}^2 \equiv a \bmod p^{n+1},$$

has to hold, we have to solve with respect to the unknown c the following congruence:

$$e_n^2 + 2ce_n p^n + c^2 p^{2n} \equiv a \bmod p^{n+1},$$

or, since $2n > n + 1$ for $n > 1$,

$$e_n^2 + 2ce_n p^n \equiv a \bmod p^{n+1}. \tag{2.7}$$

By the induction hypothesis, $e_n^2 \equiv a \bmod p^n$, that is, $a - e_n^2 \equiv 0 \bmod p^n$, or, $a - e_n^2$ is divisible by p^n. From this follows that $\frac{a - e_n^2}{p^n}$ is an integer. Then (2.7) has a solution c in integers mod p^{n+1} if and only if the equation in c:

$$\frac{a - e_n^2}{p^n} = 2e_n c \tag{2.8}$$

has a solution in integers mod p. Now, since $e_n \equiv e_1 \neq 0 \bmod p$, the inverse of $e_n \bmod p$ exists, as well as the inverse of 2, as $p \neq 2$. Thus, equation (2.8), and hence equation (2.7) too, admits the solution:

$$c \equiv \frac{1}{2e_1} \frac{a - e_n^2}{p^n} \bmod p,$$

that is, the remainder of the division of $\frac{1}{2e_1} \frac{a - e_n^2}{p^n}$ for p. With this c, which we shall call c_n, we have the new solution $e_{n+1} = e_n + c_n p^n$ in integers modulo p^{n+1}. $\diamond$

Another way to express this result is the following one.

Theorem 2.7. *In the hypothesis of Theorem 2.6 we have the series expansion*

$$\sqrt{a} = c_0 + c_1 p + c_2 p^2 + \cdots .$$

Example. Let us compute $\sqrt{2}$ with $p = 7$. Modulo 7, the equation $x^2 - 2 = 0$ has the solution $x = 3$ since $3^2 - 2 = 7 \equiv 0 \bmod 7$ (it admits another solution too, $x = 4 \equiv -3 \bmod 7$). The procedure seen in the theorem can now start: $e_1 = c_0 = 3$, and

$$c_1 = \frac{1}{2 \cdot 3} \frac{2 - 3^2}{7} = \frac{1}{6} \cdot -\frac{7}{7} = -\frac{1}{6} = -6 \equiv 1 \bmod 7,$$

so: $e_2 = c_0 + c_1 \cdot 7 = 3 + 1 \cdot 7$. Analogously:

$$c_2 = 6 \cdot \frac{2 - e_2^2}{7^2} = 6 \cdot \frac{2 - 100}{7^2} = 6 \cdot -2 \equiv 2 \bmod 7,$$

and hence $e_3 = 3 + 1 \cdot 7 + 2 \cdot 7^2$. Going on, we find $\sqrt{2} = 3.12612124 \cdots$.

In the proof of Theorem 2.6 the following algorithm is contained, where e_1 is a solution of equation mod p:

input: a, p, e_1;
output: $c_0, c_1, \ldots, c_n$ (or: $e_1, e_2, \ldots, e_{n+1}$);
$c_0 := e_1$, $m := (2e_1)^{-1} \bmod p$;
for k from 1 to n do:
$q_k := \mathrm{quotient}(a - e_k^2, p^k)$;
$c_k := q_k m \bmod p$;
$e_{k+1} := e_k + c_k \cdot p^k$.

It is immediate to see that the algorithm is completely analogous to the algorithm for rational numbers. Indeed, they are particular cases of the same algorithm which we shall see shortly. However, consider first the case $p = 2$, which we have neglected till now, and which needs a separate discussion.

Firstly, if a is odd, the equation $x^2 - a = 0$ always has a solution mod 2, i.e. $x = 1$ ($1 - a$ is even, so it is 0 mod 2). But, unlike the case $p > 2$, it has not necessarily a solution mod 4. For instance, $x^2 - 7 = 0$ has no solution in integers mod 4, as is easily verified. The reason is that $7 \not\equiv 1 \bmod 4$. Indeed, if $x^2 - a$ has a solution mod 4, this solution is either $x = 1$ or $x = 3 \equiv -1 \bmod 4$ (since a is odd). Then $1 - a \equiv 0 \bmod 4$ and $a \equiv 1 \bmod 4$; analogously, for $x = 3$ we have $a \equiv 9 \equiv 1 \bmod 4$. Conversely, if $a \equiv 1 \bmod 4$, we have the solutions $x = 1$ and $x = 3$. Hence, a necessary and sufficient condition for the equation $x^2 - a = 0$, for odd a, to have a solution in integers mod 4 is $a \equiv 1 \bmod 4$. But the equation $x^2 - a = 0$ may have a solution mod 4 even if it has no solutions mod 8. It is, for instance, the case of $x^2 - 21 = 0$. The odd numbers mod 8 are 1, 3, 5, and 7, whose squares 1, 9, 25, and 49 are congruent to 1 mod 8, while 21 is not. Thus, for the equation to have a solution it is necessary to have $a \equiv 1 \bmod 8$. Of course, this condition is also sufficient, because in this case $x = 1$ is a solution. But not only: it is also sufficient for the equation to have a solution mod 2^n for every $n \geq 3$ (so for $n = 1$ and 2 too). This is the gist of the following theorem.

Theorem 2.8. *Let $a \equiv 1 \bmod 8$. Then the equation:*

$$x^2 - a = 0$$

has a solution in integers modulo 2^n for all $n = 1, 2, \ldots$

Proof. Firstly, for $n = 1, 2, 3$ we have the solution $x = 1$. Let $n > 3$, and let e_n be a solution mod 2^n. Now we look for a solution in the form:

$$e_{n+1} = e_n + c \cdot 2^{n-1}$$

(not 2^n as in case $p \neq 2$). It must be the case that $e_{n+1}^2 \equiv a \bmod 2^{n+1}$, that is:

$$e_n^2 + 2^n e_n c + 2^{2n-2} c^2 \equiv a \bmod 2^{n+1},$$

and since $2n - 2 > n + 1$, as $n > 3$, the coefficient of c^2 is zero, and we are left with:

$$e_n^2 + 2^n e_n c \equiv a \bmod 2^{n+1}.$$

By hypothesis, $e_n^2 \equiv a \bmod 2^n$, and the solution of the previous equation is in this case the integer c such that:

$$c \equiv \frac{a - e_n^2}{2^n e_n} \bmod 2.$$

With this value of c, $e_{n+1} = e_n + c \cdot 2^{n-1}$ is the solution mod 2^{n+1}.   ◇

Example. Consider the equation $x^2 - 41 = 0$. We have the solution $x = 1$ for $n = 1, 2, 3$. A solution for $n = 4$, that is, mod 2^4, can be found as follows. With $e_3 = 1$ we have:

$$(1 + 2^2 c)^2 = 1 + 2^3 c + 2^4 c^2 \equiv 41 \bmod 2^4,$$

$$c \equiv \frac{41 - 1}{8} = 5 \equiv 1 \bmod 2,$$

so $e_4 = 1 + 1 \cdot 2^2 = 5$, and $5^2 = 25 \equiv 41 \bmod 2^4$ indeed.

We can sum up the above in the following theorem.

Theorem 2.9. *Let a be an odd number. Then:*
i)   *a is always a square modulo 2;*
ii)   *a is a square modulo 4 if and only if $a \equiv 1 \bmod 4$;*
iii)   *a is a square modulo 2^n, $n \geq 3$, if and only if $a \equiv 1 \bmod 8$.*

2.3 Newton's method

As already mentioned, series expansions yielding better and better approximations of a number are particular cases of a general method. This is *Newton's method* of successive approximations. Given a function $f(x)$ and an approximate value x_n of one of its roots, this method consists in finding a better approximation x_{n+1} by taking as x_{n+1} the intersection point of the tangent to the curve representing $f(x)$ at the point $(x_n, f(x_n))$ with the x-axis. And since this tangent has equation $y - f(x_n) = f'(x_n)(x - x_n)$, this point is

$$x_{n+1} = x_n - \frac{f(x_n)}{f'(x_n)}. \tag{2.9}$$

Now, in the case of a rational number a/b, we know that this number is root of $f(x) = bx - a$. The formula seen in Section 2.1 was, with $(b, p) = 1$, $e_{n+1} = e_n + c_n p^n$, where:

$$c_n \equiv \frac{1}{b} \frac{a - be_n}{p^n} \bmod p.$$

But $f(e_n) = be_n - a$, and $f'(x) = b$, a constant. So our c_n can be written:

$$c_n = -\frac{f(e_n)}{f'(e_n)p^n} \bmod p,$$

and hence:

$$e_{n+1} = e_n - \frac{f(e_n)}{f'(e_n)} \bmod p,$$

which is the form taken by (2.9) in the present case. Similarly, in the case of a square root $(p \neq 2)$, we have $f(x) = x^2 - a$, $f'(x) = 2x$, so

$$c_n \equiv \frac{1}{2e_n} \frac{a - e_n^2}{p^n} \bmod p,$$

or

$$c_n = -\frac{f(e_n)}{f'(e_n)p^n} \bmod p,$$

so (2.9) takes the form:

$$e_{n+1} = e_n - (\frac{f(e_n)}{f'(e_n)p^n} \bmod p)p^n.$$

This procedure holds for polynomials of any degree: if a polynomial $f(x)$ has a root $e_1 \bmod p$, and if $f'(e_1) \not\equiv 0 \bmod p^2$, then (2.9) gives a way to find a root of $f(x) \bmod p^n$ for all n. In other words, it allows us to find the p-adic expansion of a root of an arbitrary polynomial $f(x)$ with integer coefficients (within the mentioned limits). Indeed, let:

$$f(x) = \sum_{k=0}^{m} a_k x^k;$$

note that, for all integer c,

$$(x + cp)^k = \sum_{i=o}^{k} \binom{k}{i} x^{k-i}(cp)^i \equiv x^k + kcpx^{k-1} \bmod p^2,$$

since, for $i > 1$, $(cp)^i = c^i p^i \equiv 0 \bmod p^2$. From this follows:

$$f(x + cp) = \sum_{k=0}^{m} a_k(x + cp)^k \equiv \sum_{k=0}^{m} a_k x^k + cp \sum_{k=0}^{m} a_k k x^{k-1} \bmod p^2,$$

and hence:

$$f(x + cp) \equiv f(x) + cpf'(x) \bmod p^2.$$

Now, if e_1 is a root of $f(x) \bmod p$ such that $f'(e_1) \not\equiv 0 \bmod p$, for a root $e_2 \equiv e_1 + cp \bmod p^2$ the following must hold:

$$0 \equiv f(e_1 + cp) = f(e_1) + cpf'(e_1) \bmod p^2.$$

---

[2] Note that the case $f(x) = x^2 - a$ with $p = 2$ remains excluded, since $f'(x) = 2x \equiv 0 \bmod 2$.

Now, $f(e_1) \equiv 0 \bmod p$, so c solves the previous congruence mod p^2 if and only if it solves

$$cf'(e_1) \equiv -\frac{f(e_1)}{p} \bmod p.$$

But the latter admits the solution

$$c \equiv -(\frac{f(e_1)}{f'(e_1)p} \bmod p)p.$$

So we have the new approximation

$$e_2 = e_1 - (\frac{f(e_1)}{f'(e_1)p} \bmod p)p.$$

In general, in order to pass from e_n to e_{n+1}, consider the congruence:

$$f(x + cp^n) \equiv f(x) + cp^n f'(x) \bmod p^{n+1},$$

which yields the value

$$c \equiv -\frac{f(e_n)}{f'(e_n)p^n} \bmod p,$$

and the new approximation

$$e_{n+1} = e_n - (\frac{f(e_n)}{f'(e_n)p^n} \bmod p)p^n.$$

So we have the following algorithm, which subsumes the two algorithms already seen. Note that $f'(e_k) \equiv f'(e_1) \bmod p$, since e_k equals e_1 plus a multiple of p, $e_k = e_1 + tp$, and by the above for $t = c$, we have $f'(e_k) = f'(e_1 + tp) \equiv f'(e_1) \bmod p$. Set $d = \frac{1}{f'(e_1)}$, and $e_1 = e$, the starting value of the root.

input: $f(x)$, p, e, d;
output: $c_0, c_1, \ldots, c_{n-1}$ (or $e_1, e_2, \ldots, e_n$);

$e_1 := e$;
for k from 1 to $n - 1$ do:
$q_k := \text{quotient}(-f(e_k), p^k) \cdot d$;
$c_k := q_k \bmod p$;
$e_{k+1} := e_k + c_k \cdot p^k$.

Example. Let us compute some digits of the 7-adic expansion of a cubic root of -1. It is one of the three solutions modulo 7 of the equation $x^3 + 1 = 0$, that is, $x = 3$, 5 and 6. Expand the first one, $x = 3$. We have $f(x) = x^3 + 1$, $f'(x) = 3x^2$, and, with $e_1 = 3$, $f'(e_1) = f'(3) = 27 \equiv 6 \bmod 7$, and hence $d = 6$. Running the algorithm we find:

- $k = 1$: $\quad q_1 := -\frac{f(e_1)}{7} \cdot 6 = -\frac{28}{7} \cdot 6 = -\frac{168}{7} = -24$,
  $\quad\quad c_1 := -24 \bmod 7 = 4$,
  $\quad\quad e_2 := 3 + 4 \cdot 7 = 31$;

• $k = 2$: $q_2 := -\frac{f(e_2)}{7^2} \cdot 6 = -\frac{f(31)}{7^2} \cdot 6 = -608 \cdot 6 = -3648 \equiv 1 \bmod 7$,

and going on like this we find $3.4630262434\ldots \bmod 7$.

In the standard case, formula (2.9) can be used to compute a square root. Compute some digits of $\sqrt{2}$, starting with an approximate value, say $x_1 = 1$. Now, $f(x) = x^2 - 2$, $f'(x) = 2x$ and hence $f(x_1) = -1$, $f'(x_1) = 2$, and

$$x_2 = 1 - \frac{-1}{2} = \frac{3}{2} = 1.5.$$

With this value,

$$x_3 = 1.5 - \frac{2.25 - 2}{3} = \frac{4.5 - 0.25}{3} = 1.41\overline{6}.$$

So we see that the convergence to the value of $\sqrt{2}$ is in this case very fast. The usual method for computing a square root follows closely Newton's method, as seen here.

2.4 Series expansion of rational functions

A polynomial $f(x)$ with coefficients in a field (which in what follows will be the field Q of rationals) is a linear combination of the monomials $1, x, x^2, \ldots$, with coefficients equal to zero from some point on:

$$f(x) = a_0 + a_1 x + \cdots + a_n x^n + 0 \cdot x^{n+1} + \cdots .$$

In this form a polynomial is a (finite) series of powers of x, or *in base x*. If $p(x)$ is an arbitrary polynomial of first degree, we have analogously the expansion:

$$f(x) = c_0 + c_1 p(x) + c_2 p(x)^2 + \cdots + c_n p(x)^n + 0 \cdot p(x)^{n+1} + \cdots$$

in base $p(x)$. The procedure that yields such an expression is the same one used for the integers. Divide $f(x)$ by $p(x)$ (in particular, by x): $f(x) = r(x) + p(x)q(x)$, $0 \leq \partial r(x) < \partial p(x)$, and take the remainder $r(x) = c_0$, which is a constant since its degree is zero. This will be the first approximation to the polynomial: $f(x) \equiv c_0 \bmod p(x)$. Take then $\frac{f(x) - c_0}{p(x)}$, divide by $p(x)$ and take the remainder c_1, which is a constant too:

$$c_1 \equiv \frac{f(x) - c_0}{p(x)} \bmod p(x),$$

and in general:

$$c_k \equiv \frac{f(x) - (c_0 + c_1 p(x) + \cdots + c_{k-1} p(x)^{k-1})}{p(x)^k} \bmod p(x).$$

Now, if $g(x)$ is another polynomial, we may look for an analogous expansion for the rational function $f(x)/g(x)$. We have a theorem pretty analogous to Theorem 2.1.

Theorem 2.10. *Let $f(x)/g(x)$ be a rational function and $p(x)$ a first-degree polynomial with $(p(x), g(x)) = 1$. Then there exists a unique pair of polynomials $c = c(x)$ and $d(x)$, with c a constant, such that:*

$$\frac{f(x)}{g(x)} = c + \frac{d(x)}{g(x)}p(x). \tag{2.10}$$

Proof. There exist $c_0(x)$ and $d_0(x)$ such that: $g(x)c_0(x) + p(x)d_0(x) = 1$, so:

$$g(x)(c_0(x)f(x)) + p(x)(d_0(x)f(x)) = f(x).$$

Divide $c_0(x)f(x)$ by $p(x)$ to obtain $c_0(x)f(x) = p(x)q(x) + r(x)$, with $\partial r(x) = 0$. Having set:

$$c = r(x) = c_0(x)f(x) - p(x)q(x) \text{ e } d(x) = d_0(x)f(x) + q(x)g(x),$$

we have $g(x) \cdot c + p(x)d(x) = f(x)$, and hence (2.10). If $c' = $ constant, and $d'(x)$ is another pair such that $g(x)c' + p(x)d'(x) = f(x)$, subtracting form the previous formula we get $g(x)(c - c') = p(x)(d'(x) - d(x))$, and hence, since $(p(x), g(x)) = 1$, $p(x)$ divides $c - c'$. But $\partial p(x) = 1$ and $\partial(c - c') = 0$; the only possibility is that $c - c' = 0$, that is, $c' = c$, so $d'(x) = d(x)$ too. $\diamond$

Remark. Since $p(x)$ is of first degree, $p(x) = ax + b$, condition $(p(x), g(x)) = 1$ is equivalent to the condition that $-\frac{b}{a}$ is not a root of $g(x)$.

Applying this theorem to the pair $(d(x), g(x))$, we find a constant c_1 and a polynomial $d_1(x)$ and, going on, a constant c_i and a polynomial $d_i(x)$. Substituting for $d_i(x)/g(x)$ its expression:

$$\frac{d_i(x)}{g(x)} = c_{i+1} + \frac{d_{i+1}(x)}{g(x)}p(x),$$

we obtain the expansion:

$$\frac{f(x)}{g(x)} = c_0 + c_1 p(x) + \cdots + c_{n-1}p(x)^{n-1} + \frac{d_{n-1}(x)}{g(x)}p(x)^n,$$

$n = 1, 2, \ldots$ From this follows

$$f(x) = g(x)(c_0 + c_1 p(x) + \cdots + c_{n-1}p(x)^{n-1}) + d_{n-1}(x)p(x)^n.$$

The term $R_n = d_{n-1}(x)p(x)^n$ is the *order n remainder* of the expansion. From the previous formula it follows:

$$f(x) \equiv g(x)(c_0 + c_1 p(x) + \cdots + c_{n-1}p(x)^{n-1}) \text{ mod } p(x)^n, \tag{2.11}$$

$n = 1, 2, \ldots$ The infinite sequence of congruences (2.11) is usually summarised in the form

$$\frac{f(x)}{g(x)} = c_0 + c_1 p(x) + \cdots = \sum_{k \geq 0} c_k p(x)^k.$$

The computation of the coefficients c_i is performed by recursively solving congruences (2.11). To solve the first one,

$$f(x) \equiv c_0 g(x) \bmod p(x),$$

means finding c_0 in such a way that $f(x)$ and $c_0 g(x)$, divided by $p(x)$, give the same remainder. Since $p(x) = ax + b$, of first degree, this remainder is the value of $f(x)$ and $c_0 g(x)$ at $-\frac{b}{a}$. So it has to be $f(-\frac{b}{a}) = c_0 g(-\frac{b}{a})$, that is, $c_0 = f(-\frac{b}{a})/g(-\frac{b}{a})$.

For the sake of simplicity, we shall confine ourselves from now on to the polynomial $p(x) = x - a$. So:

$$c_0 = \frac{f(a)}{g(a)}.$$

To find c_1 consider, with this value of c_0, the congruence:

$$f(x) \equiv g(x)(c_0 + c_1(x - a)) \bmod (x - a)^2.$$

Since c_0 is such that $f(x) - c_0 g(x)$ is divisible by $x - a$, the quotient $q_1(x) = (f(x) - c_0 g(x))/(x - a)$ is a polynomial. Hence the previous congruence is equivalent to:

$$q_1(x) \equiv c_1 g(x) \bmod (x - a),$$

and, as above:

$$c_1 = \frac{q_1(a)}{g(a)}.$$

In general,

$$c_n = \frac{1}{g(a)} \frac{f(x) - g(x)(c_0 + c_1(x - a) + \cdots + c_{n-1}(x - a)^{n-1})}{(x - a)^n} \tag{2.12}$$

mod $(x - a)$. So we have the algorithm (for $g(a) \neq 0$):

input: $f(x)$, $g(x)$, a;
output $c_0, c_1, \ldots, c_{n-1}$ (or: $e_1, e_2, \ldots, e_n$);
$d := \frac{1}{g(a)}$;
$c_0 := f(a)d$;
$e_1 := c_0$;
for k from 1 to $n - 1$ do:
$q_k := \text{quotient}(f(x) - e_k g(x), (x - a)^k) \cdot d$;
$c_k := \text{remainder}(q_k, x - a)$;
$e_{k+1} := e_k + c_k \cdot (x - a)^k$.

Examples. 1. We have already seen that the function $\frac{1}{1-x}$ is expanded as $1 + x + x^2 + \cdots$ (as a series of powers of $x - a$ with $a = 0$), that is, with all coefficients equal to 1. Let us find again this result using the algorithm. Here $f(x) = 1$, $g(x) = 1 - x$. So $g(0) = 1$, $d = 1$, $q_0 = 1$, $c_0 = \frac{1}{1} = 1$, $e_1 = 1$.

- $k = 0$: $\quad q_1 := \text{quotient}(1 - 1 \cdot (1 - x), x) \cdot 1 = 1$,
  $\qquad\quad c_1 := 1$,
  $\qquad\quad e_2 := 1 + 1 \cdot x$;

- $k = 1$: $\quad q_2 := \text{quotient}(1 - (1 + x)(1 - x), x^2) \cdot 1$
  $\qquad\qquad = \text{quotient}(x^2, x^2) \cdot 1 = 1$,
  $\qquad\quad c_2 := 1$,
  $\qquad\quad e_2 := 1 + x + x^2$;

and so on.

2. Let us expand the function:

$$\frac{1}{1 - x - x^2},$$

and show that the coefficients of the expansion are the Fibonacci numbers $(f_0 = 0), f_1 = 1, f_2 = 1, f_3 = 2, \ldots, f_{k+2} = f_k + f_{k+1}$, for $k \geq 2$:

$$\frac{1}{1 - x - x^2} = 1 + x + 2x^2 + 3x^3 + \cdots = \sum_{k \geq 0} f_{k+1} x^k.$$

The series $\sum_{k \geq 0} f_{k+1} x^k$ is the *generating series* of the Fibonacci numbers (a power series $\sum_{k \geq 0} c_k x^k$ is called the *generating series* of the numbers c_k). From (2.12) we get $c_0 = \frac{1}{1} = 1$,

$$c_1 = \frac{1 - (1 - x - x^2)}{x} = \frac{x + x^2}{x} = 1 + x \equiv 1 \bmod x;$$

$$c_2 = \frac{1 - (1 - x - x^2)(1 + x)}{x^2} = \frac{2x^2 + x^3}{x^2} = 2 + x \equiv 2 \bmod x,$$

so $c_0 = f_1$, $c_1 = f_2$, $c_2 = f_3$. Let, by induction, $c_k = f_{k+1} = f_k + f_{k-1}$. From (2.12) we obtain:

$$c_n = \frac{1 - (1 - x - x^2) \sum_{k=0}^{n-1} f_{k+1} x^k}{x^n} \bmod x.$$

The numerator of the previous fraction is:

$$1 - \sum_{k=0}^{n-1} f_{k+1} x^k + \sum_{k=0}^{n-1} f_{k+1} x^{k+1} + \sum_{k=0}^{n-1} f_{k+1} x^{k+2} =$$

$$= (1 - f_1) + (f_1 - f_2)x + (f_1 + f_2 - f_3)x^2 + \cdots + (f_{n-2} + f_{n-1} - f_n)x^{n-1}$$
$$+ (f_n + f_{n-1})x^n + f_n x^{n+1}.$$

By induction, $f_{k+1} = f_k + f_{k-1}$ for $k < n$, so all the coefficients of the powers x^k, $k < n$, are zero. Thus, we have:

$$c_n = \frac{(f_{n-1} + f_n)x^n + f_n x^{n+1}}{x^n} \mod x,$$

so $c_n = f_n + f_{n-1}$ and hence $c_n = f_{n+1}$.

Expand now the same function by directly applying Theorem 2.10. We have:

$$(1 - x - x^2) \cdot 1 + x(1 + x) = 1, \tag{2.13}$$

and hence:

$$\frac{1}{1 - x - x^2} = 1 + \frac{1 + x}{1 - x - x^2}x. \tag{2.14}$$

Multiplying (2.13) by $1 + x$,

$$(1 - x - x^2)(1 + x) + x(1 + x)^2 = 1 + x.$$

Here $c_0(x) = 1 + x$. Dividing by x we get a quotient equal to 1 and a remainder equal to 1. So $c_1 = 1$ and $d_1(x) = (1 + x)^2 + 1 \cdot (1 - x - x^2) = 1 + 2x + x^2 + 1 - x - x^2 = 2 + x$. Then

$$\frac{1 + x}{1 - x - x^2} = 1 + \frac{2 + x}{1 - x - x^2}x$$

and, substituting in (2.14),

$$\frac{1}{1 - x - x^2} = 1 + x + \frac{2 + x}{1 - x - x^2}x^2.$$

Analogously:

$$(1 - x - x^2)(2 + x) + x(1 + x)(2 + x) = 2 + x,$$

and by the usual procedure we find $c_2 = 2$ and $d_2(x) = 3 + 2x$. From this follows:

$$\frac{1}{1 - x - x^2} = 1 + x + 2x^2 + \frac{3 + 2x}{1 - x - x^2}x^3.$$

Remark. As the last expansion shows, the coefficients c_i are exactly those of the quotient of the division of $f(x)$ by $g(x)$ ordering the two polynomials by decreasing powers of x; so the division is performed by dividing the two monomials of the lowest degree. In our case, indeed, the first quotient is 1, and the first remainder $x + x^2 = (1 + x)x = d(x) \cdot x$. The second quotient is 1 again, and the second remainder is $2x^2 + x^3 = (2 + x)x^2 = d_1(x) \cdot x^2$ and so on. It is easy to see that this is a general fact.

The algorithm suggested by Newton's method can also be applied to the case of a rational function $\frac{f(x)}{g(x)}$. As is the case for rational numbers, such a function can be seen as the solution of the equation $g(x)y - f(x) = 0$, in y with polynomials in x as coefficients. The place of the prime number p of the algorithm seen above is taken here by the variable x. Starting with a solution mod x, that is, $y = f(0)/g(0) = c_0$ (we just consider expansions in powers of x, that is, we suppose $g(0) \neq 0$), the following algorithm provides the coefficients $c_1, c_2, \ldots$ In the algorithm, d is the inverse of the derivative of the polynomial $F(y) = g(x)y - f(x)$ with respect to y, taken mod x, that is, computed at $x = 0$, or $d = \frac{1}{g(0)}$.

input: $f(x)$, $g(x)$;
output: $c_0, c_1, \ldots, c_n$ (or: $e_1, e_2, \ldots, e_n$);
$F(y) := g(x)y - f(x)$;
$d := \frac{1}{g(0)}$;
$c_0 := d$;
$e_1 := d$;
for k from 1 to $n - 1$ do:
$q_k := \text{quotient}(-F(e_k), x^k) \cdot d$;
$c_k := \text{remainder}(q_k, x)$;
$e_{k+1} := e_k + c_k \cdot x^k$).

Example. Let us expand

$$\frac{1 + x}{(1 - x)^2}.$$

Here we have $F(y) = (1 - x)^2 y - (1 + x)$, $F'(y) = (1 - x)^2$, $F'(0) = 1$, $d := 1$, $c_0 := 1$, $e_1 := 1$. Running the algorithm, we find:

- $k = 1$: $q_1 := \text{quotient}(-F(1), x)$
    $= \text{quotient}((1 - x)^2 - (1 + x), x) = -x + 3$,
    $c_1 := \text{remainder}(q_1, x) = 3$,
    $e_2 := 1 + 3x$;

- $k = 2$: $q_2 := \text{quotient}((1 - x)^2(1 + 3x) - (1 + x), x^2)$
    $= \text{quotient}(-3x^3 + 5x^2, x^2) = -3x + 5$,
    $c_2 := \text{remainder}(q_2, x) = 5$,
    $e_3 := 1 + 3x + 5x^2$,

and so on. So we obtain the generating series of odd numbers

$$1 + 3x + 5x^2 + 7x^3 + \cdots$$

Exercises

10. Prove that if $F(x) = a_0 + a_1 x + \cdots$, and if $b_k = a_0 + a_1 + \cdots + a_k$, then

$$\frac{F(x)}{1 - x} = b_0 + b_1 x + \cdots .$$

11. Determine the first digits of the expansion of the two square roots of 5 mod 11 and verify that the termwise sum gives 0 mod 11.

12. As in previous exercise, with the roots of $x^2 - 2$ mod 47.

13. Let $f(x)$ be a polynomial with non-negative integer coefficients. Let a be an integer greater than all the coefficients of $f(x)$, and let $f(a) = b$. Find the coefficients of $f(x)$.

2.5 Linear recurrence relations

We know that the series expansion of a rational number is periodic: the sequence of its digits is such that for some d we have $c_{n+d} = c_n$ starting from some n.

Let now $u_1, u_2, \ldots$ be a sequence of numbers. A relation of the form:

$$u_{n+k} = a_1 u_{n+k-1} + a_2 u_{n+k-2} + \cdots + a_{k-1} u_{n+1} + a_k u_n \qquad (2.15)$$

is called *k-th order linear recurrence relation* among the elements of the sequence. Starting from some n, the values of u_{n+k} are determined by the k preceding terms. The relation $c_{n+d} = c_n$ is of order d, with coefficients $a_{d-1} = \ldots = a_1 = 0$ and $a_d = 1$. The coefficients of a series expansion of a rational function satisfy a relation of this form, as we shall see shortly. So this is the more general notion corresponding to periodicity for rational numbers. Let us see some examples. The coefficients of $\frac{1}{1-x}$ are all equal (and equal to 1), so they satisfy the first order relation $u_{n+1} = u_n$, $n \geq 0$. The natural numbers, that is, the coefficients of $\frac{1}{(1-x)^2}$, satisfy $u_{n+2} = 2u_{n+1} - u_n$, $n \geq 0$, of order 2, and so does any other sequence of numbers in an arithmetic progression (if the difference between two consecutive terms is constant, $u_{n+2} - u_{n+1} = u_{n+1} - u_n$, then $u_{n+2} = 2u_{n+1} - u_n$). The Fibonacci numbers, as we know, satisfy $u_{n+2} = u_{n+1} + u_n$, for $n \geq 2$, also of order 2.

Theorem 2.11. *Consider the rational function* $f(x)/g(x)$, $\partial f(x) < \partial g(x)$, $g(0) \neq 0$, *and let* $\partial g(x) = k$. *Then the coefficients of the expansion:*

$$\frac{f(x)}{g(x)} = c_0 + c_1 x + c_2 x^2 + \cdots$$

satisfy a k-th order linear recurrence relation.

Proof. Let $f(x) = a_0 + a_1 x + \cdots + a_m x^m$, $g(x) = b_0 + b_1 x + \cdots + b_k x^k$. Let $n \geq m - k + 1$, and stop the expansion with the term in x^{n+k}. We have:

$$a_0 + a_1 x + \cdots + a_m x^m = (b_0 + b_1 x + \cdots + b_k x^k)(c_0 + c_1 x + \cdots + c_{n+k} x^{n+k}) + R(x).$$

The remainder $R(x)$ contains powers of x greater than $n + k$, while the left-hand polynomial only includes powers of x lower than $n+k$, since $m \leq n+k-1$; so, on the right-hand side, the coefficient of this power of x is zero. But this coefficient is $c_{n+k} b_0 + c_{n+k-1} b_1 + \cdots + c_n b_k = 0$, and hence, as $b_0 = g(0) \neq 0$,

$$c_{n+k} = -\frac{b_1}{b_0} c_{n+k-1} - \cdots - \frac{b_k}{b_0} c_n,$$

which proves the claim. $\diamond$

Remark. If $m = \partial f(x) \geq \partial g(x)$, take $n \geq m - k + 1$.

Example. With $f(x) = 1$ and $g(x) = 1 - x - x^2$ (Fibonacci), we have $b_0 = 1$, $b_1 = -1$, $b_2 = -1$, and since $k = 2$, $c_{n+2} = 1 \cdot c_{n+1} + 1 \cdot c_n$.

The converse of Theorem 2.11 holds too. In other words, Theorem 2.11 describes the only way linear recurrence relations may appear.

Theorem 2.12. *Let $u_1, u_2, \ldots$ be a sequence satisfying the linear recurrence relation $u_{n+k} = a_1 u_{n+k-1} + \cdots + a_n u_n$. Then there exist two polynomials $f(x)$ and $g(x)$ such that $\frac{f(x)}{g(x)} = u_1 + u_2 x + \cdots$.*

Proof. Let $g(x) = 1 - a_1 x - \cdots - a_k x^k$, and multiply $g(x)$ by $u_1 + u_2 x + \cdots + u_{n+1} x^n$; if m is an integer, $n \geq m \geq 1$, we have:

$$u_1 + (u_2 - a_1 u_1) x + \cdots + (u_{k+m-1} - a_1 u_{k+m-2} - \cdots - a_k u_{m-1}) x^{k+m-2}$$
$$+ (u_{k+m} - a_1 u_{k+m-1} - \cdots - a_k u_m) x^{k+m-1} +$$
$$\vdots$$
$$+ (u_{n+1} - a_1 u_n - \cdots - a_k u_{n-k+1}) x^n -$$
$$- (a_1 u_{n+1} + \cdots + a_k u_{n-k+2}) x^{n+1} + \cdots + a_k u_{n+1} x^{n+k}.$$

The coefficients of the powers of x appearing in the lines from the second to the last but one are zero by the hypothesis that the sequence (u_i) satisfies a linear recurrence. Call $f(x)$ the polynomial appearing in the first line, and $-R(x)$ the polynomial appearing in the last one. So:

$$f(x) = (1 - a_1 x - \cdots - a_k x^k)(u_1 + u_2 x + \cdots + u_{n+1} x^n) + R(x);$$

hence the u_is are indeed the coefficients of the quotient

$$\frac{f(x)}{1 - a_1 x - \cdots - a_k x^k}$$

in the division performed having ordered the two polynomials according the increasing powers of x. $\diamond$

Remark. The *characteristic equation* of relation (2.15) is the equation

$$x^k = a_1 x^{k-1} + a_2 x^{k-2} + \cdots + a_{k-1} x + a_k.$$

The roots of this equation allow us to express the generic term u_n of the sequence. Indeed, we have the following result (which we shall not prove): if $\lambda_1, \lambda_2, \ldots, \lambda_t$ are those roots, having multiplicity $m_1, m_2, \ldots, m_t$, then there exist t polynomials $p_1(x), p_2(x), \ldots, p_t(x)$, with $p_i(x)$ of degree at most m_i and such that:

$$u_n = p_1(n)\lambda_1^{n-1} + p_2(n)\lambda_2^{n-1} + \cdots + p_t(n)\lambda_t^{n-1}.$$

Moreover, if the roots λ_i are all simple, the polynomials $p_i(x)$ are constant.

Exercises

14. Prove that the sequence of squares of natural numbers satisfies the third order relation $u_{n+3} = 3u_{n+2} - 3u_{n+1} + u_n$.

Denote by s_k the sum of the first k terms of a sequence $\{u_i\}$, $i = 1, 2, \ldots$.

15. Prove that if $\{u_i\}$ is the Fibonacci sequence, then the corresponding s_k satisfy the third order relation $s_{n+3} = 2s_{n+2} - s_n$.

16. Which relation is satisfied by the terms of a geometric progression with ratio q? And what about the corresponding s_ks?

References

[C] I-5. See [CMP] for more on linear recurrence relations.

3

The resultant

Let $f = f(x)$ and $g = g(x)$ be two polynomials of degree n and m, respectively. If f and g have a non-constant factor $d = d(x)$ in common, then

$$f = dk, \qquad \partial k < n,$$
$$g = dh, \qquad \partial h < m,$$

for some polynomials k and h. From this follows

$$d = \frac{f}{k} = \frac{g}{h}$$

and

$$fh = gk. \tag{3.1}$$

Conversely, suppose that there exist polynomials k and h, with $\partial k < n$ and $\partial h < m$, such that (3.1) is satisfied. Dividing h and k by (h, k), (3.1) becomes $fh_1 = gk_1$, with $(h_1, k_1) = 1$. Then k_1 divides f, $f = k_1 w$ and hence $gk_1 = fh_1 = k_1 w h_1$, so $g = h_1 w$. Moreover, $\partial w = \partial f - \partial k_1$ and $\partial k_1 \le \partial k < \partial f$, so $\partial w > 0$. Substituting $-k$ for k, (3.1) may be rewritten as:

$$fh + gk = 0. \tag{3.2}$$

So we have found:

Theorem 3.1. *A necessary and sufficient condition for two polynomials f and g of degree n and m, respectively, to have a common non-trivial factor is the existence of two polynomials k and h, with $\partial k < n$ and $\partial h < m$ such that (3.2) holds.*

Let us see now when it is the case that two polynomials like k and h exist. Given f and g,

$$f = a_0 x^n + a_1 x^{n-1} + \cdots + a_n,$$
$$g = b_0 x^m + b_1 x^{m-1} + \cdots + b_m,$$

Machì A.: Algebra for Symbolic Computation.
DOI 10.1007/978-88-470-2397-0_3, © Springer-Verlag Italia 2012

of actual degree n and m (that is, with a_0 and b_0 different from zero), two polynomials k and h:

$$k = c_0 x^{n-1} + c_1 x^{n-2} + \cdots + c_{n-1},$$
$$h = d_0 x^{m-1} + d_1 x^{m-2} + \cdots + d_{m-1},$$

satisfying (3.2) exist if and only if all the coefficients of the polynomial $fh + gk$ are equal to zero. This leads to the following homogeneous system of $n + m$ equations in the $n + m$ unknowns $d_0, d_1, \ldots, d_{m-1}, c_0, c_1, \ldots, c_{n-1}$:

$$\begin{cases} a_0 d_0 + b_0 c_0 & = 0 \\ a_1 d_0 + a_0 d_1 + b_1 c_0 + b_0 c_1 & = 0 \\ a_2 d_0 + a_1 d_1 + a_0 d_2 + b_2 c_0 + b_1 c_1 + b_0 c_2 & = 0 \\ \cdots\cdots\cdots\cdots\cdots\cdots\cdots\cdots\cdots\cdots\cdots\cdots\cdots \\ a_n d_{m-1} + b_m c_{n-1} & = 0. \end{cases}$$

This system has a non-zero solution if and only if its matrix has determinant 0, and every non-zero solution gives rise to two polynomials k and h that satisfy (3.2). Note that, due to the structure of the system, if one of the two polynomials k and h is the zero polynomial, the other one is also zero. In other words, in a non-zero solution of the system at least one of the ds and at least one of the cs have to be different from 0. By transposing the system's matrix we get the *Sylvester matrix* $S(f, g)$ of the two polynomials:

$$S(f,g) = \begin{pmatrix} a_0 & a_1 & a_2 & \cdots & 0 & 0 & 0 \\ 0 & a_0 & a_1 & \cdots & 0 & 0 & 0 \\ \vdots & \vdots & \vdots & \ddots & \vdots & \vdots & \vdots \\ 0 & 0 & \cdots & a_0 & \cdots & a_{n-1} & a_n \\ b_0 & b_1 & b_2 & \cdots & 0 & 0 & 0 \\ 0 & b_0 & b_1 & \cdots & 0 & 0 & 0 \\ \vdots & \vdots & \vdots & \ddots & \vdots & \vdots & \vdots \\ 0 & 0 & \cdots & b_0 & \cdots & b_{m-1} & b_m \end{pmatrix}.$$

The *resultant* $R(f, g)$ of the two polynomials f and g is the determinant of the Sylvester matrix $S(f, g)$.

If f and g are polynomials in several variables, since the resultant is defined for polynomials in one variable, we may compute the resultant by considering f and g as polynomials in just one of them, and considering the remaining ones as constants. If x is the chosen variable, we shall write $R_x(f, g)$, and this resultant will be a polynomial in the remaining variables.

This discussion yields:

Theorem 3.2. *Two polynomials f and g have a common non-constant factor if and only if $R(f, g) = 0$.*

Let us see now some properties of the resultant. If one of the two polynomials is a constant, say $g = a$, set by definition:

$$R(f, a) = R(a, f) = a^{\partial f};$$

if both are constant, $f = a$ and $g = b$, then, again by definition,

$$R(f, g) = \begin{cases} 0, & \text{if } a = b = 0; \\ 1, & \text{otherwise.} \end{cases}$$

Moreover,

$$R(g, f) = (-1)^{mn} R(f, g);$$

this happens because $S(g, f)$ is obtained from $S(f, g)$ by permuting its rows $1, 2, \ldots, m-1, m, m+1, \ldots, m+n$ rearranging them to the permutation $m+1, m+2, \ldots, m+n, 1, 2, \ldots, m$, and this can be done by moving m at the end by n transpositions, and analogously for $m-1, m-2, \ldots, 1$. The total number of transpositions required is mn.

By Theorem 3.2, $\gcd(f, g) \neq 1$ if and only if $R(f, g) = 0$. The relation between the greatest common divisor of f and g and their Sylvester matrix $S(f, g)$ is explained by the fact that performing *Gaussian elimination* on the matrix $S(f, g)$ corresponds to the division algorithm to compute $\gcd(f, g)$. More precisely, let $\partial f \leq \partial g$; in the division of g by f, we multiply f by $\frac{b_0}{a_0} x^{m-n}$ and subtract the result from g. So we obtain the first partial remainder ρ_1 of the division:

$$\rho_1 = (b_1 - \frac{b_0 a_1}{a_0}) x^{m-1} + (b_2 - \frac{b_0 a_2}{a_0}) x^{m-2} + \cdots + b_m.$$

Consider now the Gaussian elimination on $S(f, g)$. Subtract from the $(m+1)$-th row (that is, from the first row of bs) the first one multiplied by b_0/a_0; the result is:

$$\begin{pmatrix} a_0 & a_1 & a_2 & \cdots & 0 \\ 0 & a_0 & a_1 & \cdots & 0 \\ \vdots & \vdots & \vdots & \ddots & \vdots \\ 0 & 0 & 0 & \cdots & a_n \\ 0 & b_1 - \frac{b_0 a_1}{a_0} & b_2 - \frac{a_2 b_0}{a_0} & \cdots & 0 \\ 0 & b_0 & b_1 & \cdots & 0 \\ \vdots & \vdots & \vdots & \ddots & \vdots \\ 0 & 0 & 0 & \cdots & b_m \end{pmatrix}.$$

In the block we obtain by deleting the first row and the first column of this matrix the elements in the first row of bs are the coefficients of ρ_1 (together with some zeros). Let us try now to eliminate the term b_0 in the first column. Subtract from the row of this b_0 the first row multiplied by $\frac{b_0}{a_0}$; the elements in the second row of bs are now the coefficients of ρ_1. Continue the operation until the term b_0 in the last row is eliminated: the result is a matrix with a number k_1 of a_0s on the main diagonal, and only zeros underneath, plus a square block. This block is $S_1 = S(f, \rho_1)$, the *Sylvester matrix of f and of the first remainder* ρ_1. Now,

$$R(f, g) = a_0^{k_1} R(f, \rho_1).$$

If $\partial f \le \partial \rho_1$, we go on dividing and find another partial remainder ρ_2. Gaussian elimination on the matrix $S(f, \rho_1)$ gives $R(f, \rho_1) = a_0^{k_2} R(f, \rho_2)$, and hence:

$$R(f, g) = a_0^{k_1 + k_2} R(f, \rho_2).$$

Keep dividing until a remainder r of degree less than the degree of f is found (so this r is *the* remainder of the division of g by f). The corresponding matrix $S(f, r)$ has dimension $n + \partial r$, and since the dimension of $S(f, g)$ is $n + m$, the number of a_0s that precede the block (f, r) is $n + m - (n + \partial r) = m - \partial r$, and we have:

$$R(f, g) = a_0^{m - \partial r} R(f, r). \tag{3.3}$$

As we have seen, the elimination operations on the matrix until now correspond to the operations performed to find the remainder r. Having obtained this remainder r, we have completed the first step of the Euclidean algorithm to divide g by f, a first step corresponding to the transition from $S(f, g)$ to $S(f, r)$. Since we have $\partial f > \partial r$, consider $S(r, f)$ and apply the previous argument; we find:

$$R(f, g) = a_0^{m - \partial r_1} (-1)^{n \cdot \partial r_1} R(r_1, r_2),$$

where we have set $r = r_1$ and where r_2 is the (final) remainder of the division of f by r_1. Finally, if the Euclidean algorithm requires k steps,

$$R(f, g) = h \cdot R(r_{k-1}, r_{k-2}),$$

where h is a constant. If f and g are relatively prime, $r_{k-1} = c$, a constant, by definition we have $R(c, r_{k-2}) = c^{\partial r_{k-2}}$. In this case $R(f, g) \ne 0$. If f and g have a common non-constant factor, then $\partial r_{k-1} \ge 1$. We have $R(r_{k-1}, r_{k-2}) = h \cdot R(r_{k-1}, \rho)$, where ρ is the last but one remainder of the division of r_{k-2} by r_{k-1}. But the remainder of this division is zero, and this implies that ρ and r_{k-1} are multiple of each other. So, the first row of $S(r_{k-1}, \rho)$ is a multiple of the $(\partial r_{k-1} + 1)$st (and the second one of the $(\partial r_{k-1} + 2)$-th and so on); thus, the determinant is zero. In this case, $R(f, g) = 0$. So we have a new proof of Theorem 3.2.

Example. Let $f = 2x^2 + 1$ and $g = 2x^5 - x^3 - 2x^2 - 2$. The division $g = fq + r$ gives as its first remainder $\rho_1 = -2x^3 - 2x^2 - 2$, as its second $\rho_2 = -2x^2 + x - 2$, and finally $\rho_3 = r = x - 1$. The matrix $S(f, g)$ is

$$\begin{pmatrix} 2 & 0 & 1 & 0 & 0 & 0 & 0 \\ 0 & 2 & 0 & 1 & 0 & 0 & 0 \\ 0 & 0 & 2 & 0 & 1 & 0 & 0 \\ 0 & 0 & 0 & 2 & 0 & 1 & 0 \\ 0 & 0 & 0 & 0 & 2 & 0 & 1 \\ 2 & 0 & -1 & -2 & 0 & -2 & 0 \\ 0 & 2 & 0 & -1 & -2 & 0 & -2 \end{pmatrix}.$$

Use Gaussian elimination. Sum to the sixth row the first one multiplied by -1, and in the matrix so obtained subtract the second row from the last one. The result is:

$$\begin{pmatrix} 2 & 0 & 1 & 0 & 0 & 0 & 0 \\ 0 & 2 & 0 & 1 & 0 & 0 & 0 \\ 0 & 0 & 2 & 0 & 1 & 0 & 0 \\ 0 & 0 & 0 & 2 & 0 & 1 & 0 \\ 0 & 0 & 0 & 0 & 2 & 0 & 1 \\ 0 & 0 & -2 & -2 & 0 & -2 & 0 \\ 0 & 0 & 0 & -2 & -2 & 0 & -2 \end{pmatrix}.$$

The bottom-right 5×5 block is $S_1 = S(f, \rho_1)$. Now, $R(f, g) = 2^2 \cdot R(f, \rho_1)$. Gauss elimination goes on now in the matrix S_1. In S_1, sum the first row to the fourth one and in the resulting matrix sum the second row to the last one. We get:

$$\begin{pmatrix} 2 & 0 & 1 & 0 & 0 \\ 0 & 2 & 0 & 1 & 0 \\ 0 & 0 & 2 & 0 & 1 \\ 0 & -2 & 1 & -2 & 0 \\ 0 & 0 & -2 & 1 & -2 \end{pmatrix}.$$

The bottom-right 4×4 block is $S_2 = S(f, \rho_2)$. Here $R(f, g) = 2^3 \cdot R(f, \rho_2)$. In S_2, sum the third row to the first one; the new matrix is:

$$\begin{pmatrix} 2 & 0 & 1 & 0 \\ 0 & 2 & 0 & 1 \\ 0 & 1 & -1 & 0 \\ 0 & 0 & 1 & -1 \end{pmatrix},$$

and the bottom-right 3×3 block is $S_3 = S(f, \rho_3)$. Hence,

$$R(f, g) = 2^4 \cdot R(f, r).$$

(Note that the exponent 4 is equal to $\partial f - \partial r = 5 - 1$.)

The first step in Euclidean algorithm corresponds to the elimination in $S(f, g)$, and here it leads to $S_3 = S(f, r)$. Go on with the algorithm, and divide f by r, obtaining a first partial remainder $\rho_4 = 2x + 1$ and a second (and last) $\rho_5 = 3$. Since $\partial f > \partial r$, consider the matrix $S_3' = S(r, f)$, obtained from S_3 by transposing the first and second rows, and then the second and third ones (so the determinant does not change):

$$\begin{pmatrix} 1 & -1 & 0 \\ 0 & 1 & -1 \\ 2 & 0 & 1 \end{pmatrix}.$$

Add the first row multiplied by -2 to the last one; we get

$$\begin{pmatrix} 1 & -1 & 0 \\ 0 & 1 & -1 \\ 0 & 2 & 1 \end{pmatrix},$$

hence:

$$S_4 = \begin{pmatrix} 1 & -1 \\ 2 & 1 \end{pmatrix},$$

which is $S(r, \rho_4)$, and whose determinant is equal to 3. By definition, 3 is the resultant $R(x - 1, 3) = R(r, \rho_5)$.

To sum up:

$$R(f, g) = 2^2 \cdot R(f, \rho_1),$$
$$R(f, \rho_1) = 2 \cdot R(f, \rho_2),$$
$$R(f, \rho_2) = 2 \cdot R(f, r) = R(r, f),$$
$$R(r, f) = R(r, \rho_4) = R(r, \rho_5),$$

hence $R(f, g) = 2^4 \cdot 3 = 48$. We may conclude that f and g are relatively prime.

Remark. The number of steps required to obtain S_{i+1} from S_i in the elimination procedure is equal to the smaller of the degrees of the two polynomials S_{i+1} is the Sylvester matrix of.

Formula (3.3) suggests a way to compute recursively the resultant, and hence the following algorithm:

input: f, g;
output: R, $n = \partial f$, $m = \partial g$;
if $n > m$ then $R := (-1)^{mn} R(g, f)$, else do:
$a_n :=$ leadingcoeff(f);
if $n = 0$ then $R = a_n^m$, else do:
$r :=$ remainder(g, f);
if $r = 0$ then $R := 0$, else do:
$p := \partial r$;
$R := a_n^{m-p} R(f, r)$.

The following theorem provides an expression for the resultant in terms of the roots of the two polynomials. We need a lemma first.

Lemma 3.1. *Let* $\alpha_1, \alpha_2, \ldots, \alpha_n$ *and* $\beta_1, \beta_2, \ldots, \beta_m$ *be the roots of* $f(x)$ *and* $g(x)$, *respectively, in an extension of the field of the coefficients. Then the following equalities hold:*

$$a_0^m \prod_{f(x)=0} g(x) = (-1)^{mn} b_0^n \prod_{g(x)=0} f(x) = a_0^m b_0^n \prod_{i,j} (\alpha_i - \beta_j). \qquad (3.4)$$

Proof. (The expression $\prod_{f(x)=0} g(x)$ means $\prod_{i=1}^n g(\alpha_i)$, that is, the product of the values $g(x)$ takes on the roots of $f(x)$.) Let us prove that the first and the second quantities are equal to the third one. For the first, we have

$$g(x) = b_0(x - \beta_1)(x - \beta_2) \cdots (x - \beta_m),$$

so $g(\alpha_i) = b_0(\alpha_i - \beta_1)(\alpha_i - \beta_2) \cdots (\alpha_i - \beta_m)$.

The product of all these expressions for $i = 1, 2, \ldots, n$ is

$$\prod_{f(x)=0} g(x) = b_0^n \prod_{i,j} (\alpha_i - \beta_j).$$

For the second, $f(x) = a_0(x - \alpha_1)(x - \alpha_2) \cdots (x - \alpha_n)$, and hence $f(\beta_j) = a_0(\beta_j - \alpha_1)(\beta_j - \alpha_2) \cdots (\beta_j - \alpha_n)$, or $(-1)^n f(\beta_j) = a_0(\alpha_1 - \beta_j)(\alpha_2 - \beta_j) \cdots (\alpha_n - \beta_j)$. The product of all these expressions for $j = 1, 2, \ldots, m$ is

$$(-1)^{mn} \prod_{g(x)=0} f(x) = a_0^m \prod_{i,j} (\alpha_i - \beta_j),$$

hence the result.   ◇

Theorem 3.3. *The following equality holds, where* $m = \partial g$:

$$R(f, g) = a_0^m \prod_{f(x)=0} g(x).$$

Proof. Let $\partial f \le \partial g$. Proceed by induction on ∂f, the degree of the first polynomial of the pair f, g. If $\partial f = 0$, then $f = a_0$, a constant, and by definition $R(a_0, g) = a_0^m$. Let $\partial f > 0$. If $g = fq + r$ we have (cf. formula (3.3)):

$$R(f, g) = a_0^{m - \partial r} R(f, r).$$

Since $\partial f > \partial r$, consider $R(r, f)$. By the induction hypothesis:

$$R(r, f) = r_0^n \prod_{r(x)=0} f(x) = (-1)^{n \partial r} a_0^{\partial r} \prod_{f(x)=0} r(x),$$

where the second equality follows from the first one in (3.4). But if we compute $g = fq + r$ at a root α of f we have $g(\alpha) = r(\alpha)$, and hence $\prod_{f(x)=0} r(x) =$

$\prod_{f(x)=0} g(x)$. Then:

$$R(f,g) = a_0^{m-\partial r} R(f,r) = a_0^{m-\partial r}(-1)^{n\partial r} R(r,f)$$
$$= a_0^{m-\partial r}(-1)^{n\partial r}(-1)^{n\partial r} a_0^{\partial r} \prod_{f(x)=0} g(x)$$
$$= a_0^m \prod_{f(x)=0} g(x).$$

If $\partial f > \partial g$, consider $R(f,g) = (-1)^{mn} R(g,f)$. The argument above yields $R(g,f) = b_0^n \prod_{g(x)=0} f(x)$, and the result follows from the first equality in (3.4). $\diamond$

Corollary 3.1. *Let f, g and h be three polynomials. Then:*

$$R(fg,h) = R(f,h)R(g,h).$$

Proof. By the theorem, $R(fg,h) = (a_0 b_0)^{\partial h} \prod_{fg=0} h(x)$. But $fg(x) = f(x)g(x) = 0$ if and only if $f(x) = 0$ or $g(x) = 0$. From this follows $\prod_{fg(x)=0} h(x) = \prod_{f(x)=0} h(x) \cdot \prod_{g(x)=0} h(x)$, and we have the result. $\diamond$

Examples. 1. Let $z = a+bi$ be a complex number, $\bar{z} = a-bi$ its conjugate. These two numbers are the values of the polynomial $a+bx$ at the points i and $-i$, respectively, that is, at the roots of $x^2 + 1$. Their product is $a^2 + b^2$, the *norm* of z, and hence $N = R(x^2+1, a+bx)$. This can be verified by computing the determinant of the matrix: $\begin{pmatrix} b & a & 0 \\ 0 & b & a \\ 1 & 0 & 1 \end{pmatrix}$ whose value is indeed $a^2 + b^2$. The norm of the complex number $a + bi$ can also be defined as the norm of the polynomial $a + bx$ with respect to the roots of $x^2 + 1$. More in general, we may define the *norm of a polynomial g with respect to the roots of another polynomial f* as the resultant $R(f,g)$.

2. Let us prove that if A is a matrix and f a polynomial, then $\det(f(A)) = R(f,p(x))$, where $p(x) = \det(A - xI)$ is the characteristic polynomial of A. Let $\alpha_1, \alpha_2, \ldots, \alpha_m$ be the roots of $f(x)$. Then $p(\alpha_i) = \det(A - \alpha_i I)$. By definition, $R(f,p(x)) = \prod_{i=1}^m p(\alpha_i)$; from $f(x) = \prod_{i=1}^m (x - \alpha_i)$, we get $f(A) = \prod_{i=1}^m (A - \alpha_i I)$, and hence $\det(f(A)) = \prod_{i=1}^m \det(A - \alpha_i I) = \prod_{i=1}^m p(\alpha_i)$. From this follows that $\det(f(A)) = R(f,p(x))$. In particular, $\det(A) = R(x,p(x))$.

Exercises

1. With $m = \partial g$, we have $R(ax+b,g) = a^m g(-\frac{b}{a})$. In particular, $R(x-a,g) = g(a)$.

2. Prove Theorem 3.3 using the previous exercise and Corollary 3.1. [*Hint*: by induction on ∂f, the degree of the first polynomial.]

3. Let y be a variable. Prove that, with $m = \partial g$, we have $R_x(f, g - y) = a_0^m \prod_{f(x)=0} (g(x) - y)$. (For $y = 0$ we obtain Theorem 3.3. $R(f, g)$ is the constant term of $R_x(f, g - y)$, which is a polynomial in y.)

4. If $f(x)$, $g(x)$ and $h(y)$ are three polynomials, we have $R_y(f(h), g(h)) = (h_0^{nm} R_x(f, g))^{\partial h}$, except the case in which $h(y)$ is a constant such that $f(h) = g(h) = 0$.

5. (A characterisation of the resultant) Let F be a function that associates with every pair of polynomials over a field K an element of K in such a way that:

i) $F(f, 0) = 0$, if f is not a constant;

ii) $F(g, f) = (-1)^{mn} F(f, g)$;

iii) if $\partial f \leq \partial g$ and $g = fq + r$, then $F(f, g) = a_0^{m-\partial r} F(f, r)$, where a_0 is the leading coefficient of f.

Prove that $F(f, g)$ is the resultant $R(f, g)$. [*Hint:* by induction over $\min(\partial f, \partial g)$.]

6. Let P_i be the vector space of polynomials of degree at most i, and let f and g be two polynomials of degrees n and m, respectively. Let $P_m \times P_n$ be the vector space of pairs of polynomials (p, q) and consider the linear map $\phi : P_m \times P_n \to P_{m+n}$ given by $(p, q) \to fp + gq$. Prove that the matrix of ϕ is Sylvester matrix $S(f, g)$.

7. Prove that there exist two polynomials $a(x)$ and $b(x)$ such that:

$$R(f, g) = a(x)f(x) + b(x)g(x),$$

and that if f and g have integer coefficients then a and b have too.

8. Let R_i be the determinant obtained from $R(f, g)$ by deleting the first and last i columns, the first i rows of those corresponding to f and the first i rows of those corresponding to g. Prove that a necessary and sufficient condition for f and g to have a gcd of degree d is $R = 0$, $R_1 = 0$, ..., $R_{d-1} = 0$, $R_d \neq 0$.

3.1 Applications

The resultant allows us to solve the following problem: given a polynomial f, find a polynomial g whose roots are a given rational function of the roots of f. Of course, this problem is interesting only when the roots of f are unknown.

First of all, a polynomial f can be expressed as the resultant of two polynomials. Indeed, let $\alpha_1, \alpha_2, \ldots, \alpha_n$ be the roots of f (in an extension of the field); then:

$$f(x) = a_0(x - \alpha_1)(x - \alpha_2) \cdots (x - \alpha_n) = a_0 \prod_{f(y)=0} (x - y),$$

and hence:

$$f(x) = R_y(f(y), -y + x),$$

which is the required expression. The problem mentioned above is solved here in the case in which the rational function is the identity function.

Suppose now that the rational function is a polynomial $p(x)$. The polynomial $\prod_{f(y)=0}(x-p(y))$ has as its roots $p(\alpha_i)$, so it is a solution of our problem; hence, let us take

$$g(x) = R_y(f(y), x - p(y)).$$

Let now $p(x)/q(x)$ be an arbitrary rational function. Of course, for our problem to admit a solution it is necessary that $q(x)$ is never zero at a root of $f(x)$, so q and f cannot have common roots. Let

$$g_1(x) = \prod_{f(y)=0} \left(x - \frac{p(y)}{q(y)}\right) = \prod_{f(y)=0} (q(y)x - p(y)) \cdot \frac{1}{\prod_{f(y)=0} q(y)}.$$

Then $g_1(x)$ is a solution and $\prod_{f(y)=0}(q(y)x - p(y))$ too, and hence also:

$$g(x) = R_y(f(y), q(y)x - p(y)).$$

Note that for $q(y) = 1$ we find again the previous formula for the case of a polynomial.

Remark. We may always substitute a polynomial function for a rational function of the roots. In other words, if $p(\alpha)/q(\alpha)$ is a rational function of the root α of $f(x)$, then there is a polynomial $r(x)$ such that $p(\alpha)/q(\alpha) = r(\alpha)$ (the polynomial $r(x)$ being the same for all roots of $f(x)$). Indeed, since f and q are coprime (if they had a common factor in the field, they would have a common root in an extension of the field), we have $af + bq = 1$ and hence, multiplying by p we have $a_1 f + rq = p$, for some a_1 and r. Computing at α we obtain $r(\alpha)q(\alpha) = p(\alpha)$, since $f(\alpha) = 0$, so r is the polynomial we were looking for. Note that the reduction to r is performed by rational operations in the coefficients of p and q, so r has coefficients in the same field as p and q. Moreover, if $\partial r \geq \partial f$, by dividing we get $r = fq_1 + r_1$, with $\partial r_1 < \partial f$, and, computing at α, $r(\alpha) = r_1(\alpha)$. That is, we may obtain a reduction of the rational function to a polynomial of degree less than that of $f(x)$.

Let us consider now some particular cases of the function $p(x)/q(x)$.

1. *Augmented roots.* The polynomial $g(x)$ is required to have as its roots $\alpha + k$, where k is a number and α a root of $f(x)$. The solution is, of course, $g(x) = f(x - k)$ which, expressed as a resultant, is:

$$g(x) = R_y(f(y), x - y - k). \tag{3.5}$$

For some values of k the polynomial g is simpler than f, so, a priori, it is easier to find a root β of g; then, a root of f is simply $\alpha = \beta - k$. For instance, it is possible to determine g in such a way that the term in x^{n-1} is missing: in (3.5) the coefficient of x^{n-1} is $a_1 - na_0 k$, so it suffices to choose $k = a_1/na_0$. (The fact that the coefficient of x^{n-1} in g is $a_1 - na_0 k$ can also be seen from the fact that $g(x)$ is equal to $f(x-k) = a_0(x-k)^n + a_1(x-k)^{n-1} + \cdots + a_n$, and applying the binomial theorem.) In the special case of a quadratic equation, all of the above leads to the well-known quadratic formula. Let $f(x) = a_0 x^2 + a_1 x + a_2$;

then

$$g(x) = R_y(f(y), -y + x - k) = \begin{vmatrix} a_0 & a_1 & a_2 \\ -1 & x - k & 0 \\ 0 & -1 & x - k \end{vmatrix}$$

and hence:

$$g(x) = a_0 x^2 + (a_1 - 2ka_0)x + a_0 k^2 - a_1 k + a_2.$$

Choosing $k = a_1/2a_0$, the term in x^{n-1}, that is, in this case, the first degree term, vanishes and we have:

$$g(x) = a_0 x^2 + \frac{a_1^2}{4a_0} - \frac{a_1^2}{2a_0} + a_2 = a_0 x^2 - \frac{a_1^2 - 4a_0 a_2}{4a_0}.$$

The roots of $g(x)$ can be obtained by a simple square root extraction: $\beta = \pm\sqrt{(a_1^2 - 4a_0 a_2)/4a_0^2}$, and those of $f(x)$ are:

$$\alpha = \beta - k = \frac{\pm\sqrt{a_1^2 - 4a_0 a_2}}{2a_0} - \frac{a_1}{2a_0} = \frac{-a_1 \pm \sqrt{a_2^2 - 4a_0 a_2}}{2a_0}.$$

In the same way we can reduce a third degree polynomial to the form $x^3 + px + q$.

2. *Multiples of roots.* If $g(x)$ is assigned $k\alpha_i$ as roots, for some k, then:

$$g(x) = R_y(f(y), x - ky)$$

and $g(x) = a_0 x^n + a_1 k x^{n-1} + a_{n-2} k^2 x^{n-2} + \cdots + a_n x^n$. (Hence $g(k\alpha) = k^n f(\alpha) = 0$.) In particular, for $k = -1$ we have:

$$g(x) = \sum_{i=0}^{n} a_{n-i}(-1)^i x^{n-i},$$

which can be obtained by alternately changing sign to the coefficients of $f(x)$.

3. *Reciprocal roots.* If α_i are the roots of $f(x)$, we require $g(x)$ to have $1/\alpha_i$ as roots. Hence, $p(x)/q(x) = 1/x$ and

$$g(x) = R_y(f(y), xy - 1).$$

The Sylvester matrix is:

$$\begin{pmatrix} a_0 & a_1 & a_2 & \cdots & a_{n-1} & a_n \\ x & -1 & 0 & \cdots & 0 & 0 \\ 0 & x & -1 & \cdots & 0 & 0 \\ \vdots & \vdots & \vdots & \ddots & \vdots & \vdots \\ 0 & 0 & 0 & \cdots & x & -1 \end{pmatrix},$$

whose determinant, up to a factor $(-1)^n$, is $a_n x^n + a_{n-1} x^{n-1} + \ldots + a_0$, which is the required polynomial g. Its coefficients are the same as those of f but in reverse order, and $g(x) = x^n f(1/x)$.

4. *Linear fractional transformations.* Let the rational function have the form:

$$\frac{p(x)}{q(x)} = \frac{ax + b}{cx + d}.$$

The three cases $x + k$, kx and $1/x$ are particular cases of this one. However, a linear fractional transformation can always be reduced to a sequence of transformations of those three kinds. Indeed, if $c \neq 0$,

$$x \to x + \frac{d}{c} = u \to \frac{1}{u} = v \to \frac{bc - ad}{c^2} v = w \to w + \frac{a}{c} = \frac{ax + b}{cx + d}.$$

This entails computing four resultants:

$$g_1(x) = R_y(f(y), x - y - \frac{d}{c}),$$

$$g_2(x) = R_y(g_1(y), xy - 1),$$

$$g_3(x) = R_y(g_2(y), x - \frac{bc - ad}{c^2} y),$$

$$g(x) = R_y(g_3(y), x - y - \frac{a}{c}),$$

and the last one, $g(x)$, is the polynomial whose roots are $(a\alpha + b)/(c\alpha + d)$. If, on the other hand, $c = 0$, then the sequence of transformations reduces to just two of them:

$$x \to \frac{a}{d} x = u \to u + \frac{b}{d} = \frac{ax + b}{d},$$

and analogously for the sequence of resultants:

$$g_1(x) = R_y(f(y), x - \frac{a}{d} y),$$

$$g(x) = R_y(g_1(y), x - y - \frac{b}{d}).$$

5. *Invariance under transformations.* In which case does the polynomial g, obtained from f with a rational transformation, have the same roots as f? For this to happen, it is clearly necessary and sufficient for g to be a scalar multiple of f. Consider two cases:

1. Let $k = -1$ in case 2; that is, we want g to have the roots $-\alpha_i$ (opposite roots). Then g is equivalent to f if and only if $a_1 = a_3 = \ldots = 0$. If n is even, $n = 2m$, then we have $g(x) = a_0 x^{2m} + a_2 x^{2m-2} + \cdots + a_{2m}$; if n is odd, $n = 2m + 1$, then $g(x) = a_0 x^{2m+1} + a_2 x^{2m-2} + \cdots + a_{2m} x$. The latter expression reduces to the former if we take into account the root $\alpha = 0$ (which equals its opposite $-\alpha$), and by setting $y = x^2$ we have a polynomial whose degree is one half of the degree of f.

2. Consider the reciprocal transformation seen in case 3; the coefficients of g are those of f but in reverse order. So we find $\rho a_i = a_{n-i}$, $i = 1, 2, \ldots, n$, for some ρ, and from $\rho a_0 = a_n$ and $\rho a_n = a_0$ it follows $\rho^2 = 1$ and $\rho = \pm 1$. Then, for all i, either $a_i = a_{n-i}$ or $a_i = -a_{n-i}$.

6. *Eliminating a variable.* Let f and g be two polynomials in x and y:

$$f(x, y) = a_0(y)x^n + a_1(y)x^{n-1} + \cdots + a_n(y),$$
$$g(x, y) = b_0(y)x^m + b_1(y)x^{m-1} + \cdots + b_m(y),$$

written as polynomials in x, with polynomials in y as coefficients. We are looking for a common root of f and g. Consider the resultant $R_x(f, g)$; this is a polynomial in y, say $r(y)$. If $x = \alpha$ and $y = \beta$ is a common root of f and g, the two polynomials in x, $f(x, \beta)$ and $g(x, \beta)$ have the common root α, so their resultant, $r(\beta)$, has to be zero. From this follows that β is a root of $r(y)$.

Conversely, if $r(\beta) = 0$ for some β, then the resultant $r(y)$ admits the root β, so the two polynomials $f(x, \beta)$ and $g(x, \beta)$ have a common root (or their leading coefficients are zero: $a_0(\beta) = b_0(\beta) = 0$). Thus the problem of finding a root common to two polynomials in two variables, f and g, reduces to that of finding a root of the polynomial $r(y)$ in the single variable y: the variable x has been eliminated.

Remark. As remarked by G. Ascoli, in the common parlance, the phrase "eliminating a variable x" between two equations $f = 0$ and $g = 0$ is often used with the rather vague meaning "deducing from the two equations a third one not containing x". The precise meaning of the phrase is instead "finding the necessary and sufficient condition the other variables have to satisfy for the polynomials f and g, considered as polynomials in x, to have a common root". And this condition, by the above, is for the resultant $R_x(f, g)$ to equal zero.

Examples. 1. Find the common roots of the polynomials[1]:

$$f(x, y) = x^2 y + 3xy + 2y + 3,$$
$$g(x, y) = 2xy - 2x + 2y + 3.$$

Eliminate x. Write the two equations according the increasing powers of x:

$$f(x, y) = y \cdot x^2 + 3y \cdot x + (2y + 3),$$
$$g(x, y) = (2y - 2) \cdot x + (2y + 3).$$

Hence:

$$R_x(f, g) = \begin{vmatrix} y & 3y & 2y + 3 \\ 2y - 2 & 2y + 3 & 0 \\ 0 & 2y - 2 & 2y + 3 \end{vmatrix} = 2y^2 + 11y + 12.$$

---

[1] This example and the following one are from [Ku], page 319.

The resultant's roots are $\beta_1 = -4$ and $\beta_2 = -\frac{3}{2}$, and for these values the leading coefficients of f and g are not zero. Thus each of these values, together with a corresponding value of x, yields a common root of f and g. We have:

$$f(x, -4) = -4x^2 - 12x - 5,$$
$$g(x, -4) = -10x - 5,$$

admitting the common root $\alpha_1 = -\frac{1}{2}$. For the other value,

$$f(x, -\frac{3}{2}) = -\frac{3}{2}x^2 - \frac{9}{2}x,$$
$$g(x, -\frac{3}{2}) = -5x,$$

and $\alpha_2 = 0$ is a common root. In conclusion, $(-\frac{1}{2}, -4)$ and $(0, -\frac{3}{2})$ are the solutions to our problem.

The example that follows shows that things are not always so simple.

2. Let:

$$f(x, y) = 2x^3 y - xy^2 + x + 5,$$
$$g(x, y) = x^2 y^2 + 2xy^2 - 5y + 1,$$

which we write as polynomials in y, the variable we want to eliminate (because f and g are of degree 2 in y, and f is of degree 3 in x):

$$f(x, y) = (-x) \cdot y^2 + (2x^3) \cdot y + (x + 5),$$
$$g(x, y) = (x^2 + 2x) \cdot y^2 - 5y + 1.$$

The resultant is:

$$R_y(f, g) = 4x^8 + 8x^7 + 11x^6 + 84x^5 + 161x^4 + 154x^3 + 96x^2 - 125x,$$

which admits the root $x = 0$. For this values, the leading coefficients of f and g become zero. Moreover, $f(0, y)$ and $g(0, y)$ have no common roots. The other roots of the resultant are not easy to compute. In any case, for none of them the leading coefficients of f and g are simultaneously zero, so every non-zero root of the resultant, together with the corresponding value of x, gives a common root of f and g.

7. *The discriminant.* Consider now the multiple roots of a polynomial f. We know that α is a multiple root of f if and only if α is also a root of $f'(x)$, the derivative of f. Therefore, $R(f, f') = 0$ *is a necessary and sufficient condition for f to have a multiple root.* By Lemma 3.1,

$$R(f, f') = a_0^{n-1} \prod_{i=1}^{n} f'(\alpha_i),$$

and differentiating the equality $f(x) = a_0 \prod_{k=1}^{n}(x - \alpha_k)$ we have:

$$f'(x) = a_0 \sum_{k=1}^{n} \prod_{j \neq k}(x - \alpha_j).$$

For $x = \alpha_i$, all the summands become zero except the i-th; from this follows

$$f'(\alpha_i) = a_0 \prod_{j \neq i}(\alpha_i - \alpha_j),$$

and

$$R(f, f') = a_0^{n-1} a_0^n \prod_{i=1}^{n} \prod_{j \neq i}(\alpha_i - \alpha_j).$$

Two factors $\alpha_i - \alpha_j$ and $\alpha_j - \alpha_i$ correspond to each pair of indices (i, j), $i > j$, their product being $-(\alpha_i - \alpha_j)^2$. Since there are $n(n-1)/2$ pairs of this kind, we have:

$$R(f, f') = (-1)^{\frac{n(n-1)}{2}} a_0^{2n-1} \prod_{i<j}(\alpha_i - \alpha_j)^2$$

$$= (-1)^{\frac{n(n-1)}{2}} a_0 D,$$

where

$$D = a_0^{2n-2} \prod_{i<j}(\alpha_i - \alpha_j)^2$$

is the *discriminant* of f. Thus, up to sign, the discriminant of f is the resultant $R(f, f')$ divided by the leading coefficient of f.

Example. Consider a second degree polynomial, $f = ax^2 + bx + c$. Since $f' = 2ax + b$, we have:

$$R(f, f') = \begin{vmatrix} a & b & c \\ 2a & b & 0 \\ 0 & 2a & b \end{vmatrix} = a(-b^2 + 4ac).$$

Here $n(n-1)/2 = 1$ and hence $D = -a^{-1}R(f, f') = b^2 - 4ac$, as is well known from elementary algebra. In an analogous way, it can be proved that the discriminant of the reduced third degree equation $x^3 + px + q$ is $D = -4p^3 - 27q^2$.

Exercises

10. Prove the following property of the discriminant:

$$D(fg) = D(f)D(g)[R(f, g)]^2.$$

11. Determine the discriminant of the polynomials $x^{n-1} + x^{n-2} + \cdots + x + 1$ and $x^n + a$.

12. Prove that $D(x^n + a) = (-1)^{n(n-1)/2} n^n a^{n-1}$.

13. Prove that if $f(x)$ has roots α_i and $g(x)$ has roots β_j, then the polynomial:

$$R_1(x) = R_y(f(y), g(x - y))$$

has roots $\alpha_i + \beta_j$. If $h(x)$ is another polynomial with roots γ_k, then the polynomial:

$$R_2(x) = R_z(R_1(x), h(x - z))$$

has roots $\alpha_i + \beta_j + \gamma_k$.

8. Hurwitz polynomials. Let α_i, $i = 1, \ldots, n$, be the roots of the polynomial $f(x)$, and let y be an indeterminate. The polynomials $f(y - x)$ and $f(y + x)$ in the variable x have as roots $y - \alpha_i$ and $-y + \alpha_i$, respectively. If these two polynomials have a common root, then $y - \alpha_i = -y + \alpha_j$, for some i and j, and hence

$$y = \frac{\alpha_j + \alpha_i}{2}.$$

Conversely, if y has this form, then $y - \alpha_i = -y + \alpha_j$ is a common root of $f(y - x)$ and $f(y + x)$. From this it follows that the polynomial in y:

$$r(y) = R_x(f(y - x), f(y + x)) \tag{3.6}$$

of degree n^2, where n is the degree of f, has as its roots the half-sums of the roots of $f(x)$. The equation $r(y) = 0$ is sometimes called *half-sum equation*.

A polynomial f with real coefficients is a *Hurwitz polynomial* if its real roots are negative and the complex ones have negative real parts. The irreducible factors of a polynomial with real coefficients are either linear, of the form $x - \alpha$, in the case of a real root α, or quadratic, of the form $x^2 - (\alpha + \bar{\alpha})x + \alpha\bar{\alpha}$, in the case of a complex root α. If α is a negative real root, then $x - \alpha$ has positive coefficients; if α is complex with negative real part, then the sum $\alpha + \bar{\alpha}$ is negative, so the coefficient $-(\alpha + \bar{\alpha})$ is positive, and so is $\alpha\bar{\alpha}$, which is a sum of squares. So all the coefficients of the irreducible factors of $f(x)$ are positive, and f, as a product of factors with positive coefficients, is a polynomial all of whose coefficients have the same sign, that is, the sign of the leading coefficient. We have the following theorem.

Theorem 3.4. *Let f be a polynomial with real coefficients all of the same sign. Hence, f is a Hurwitz polynomial if and only if all the coefficients of the polynomial $r(y)$ in (3.6) have the same sign.*

Proof. If f is Hurwitz, every root has negative real part, and the same holds for the half-sum of any two roots. So, by the above, $r(y)$ is Hurwitz too, and all its coefficients have the same sign. Conversely, if $r(y)$ has coefficients all of the same sign, the real roots are necessarily negative. But these real roots are half-sums of pairs of conjugate roots of $f(x)$, that is, are the real parts of the complex roots of $f(x)$, so the complex roots of $f(x)$ have negative real part. $\Diamond$

Remark. A Hurwitz polynomial is also called *stable*. This naming comes from the theory of differential equations. For a physical system to be asymptotically stable in a neighbourhood of an equilibrium point, the condition $\lim_{t\to\infty} e^{\lambda t} = 0$ has to hold, where λ is an arbitrary root of the polynomial of degree n associated with the differential equation of order n with constant coefficients that describes the physical phenomenon. If $\lambda = \alpha + i\beta$, then $e^{\lambda t} = e^{\alpha t}e^{i\beta t} = e^{\alpha t}(\cos(\beta t) + i\sin(\beta t))$. Therefore, the dominating term is $e^{\alpha t}$, and the condition for the limit to be 0 is equivalent to the condition $\alpha < 0$.

9. *Square of roots.* A polynomial $g(x)$ whose roots are the squares of $f(x)$ is

$$g(x) = R_y(f(y), x - y^2).$$

Let us see two applications.

9a. *Upper bound for the number of real roots.* By Descartes' rule of signs, the number of positive real roots of a polynomial $f(x)$ with real coefficients is less or equal than the number N of sign changes in $f(x)$ (it equals N minus an even number). Now, the square of a real number is positive, so the number of real roots of $f(x)$ cannot be greater than the number of positive roots of $g(x)$.

Example. Let $f(x) = x^5 + x^3 + x^2 + 2x + 3$. Here $g(x) = x^5 + 2x^4 + 5x^3 + 3x^2 - 2x - 9$, and hence $N = 1$. We may conclude that $f(x)$ has four complex roots.

9b. *Graeffe's method.* This method is used to compute approximate roots of polynomials. The main remark is the following one. Let $\alpha_1, \alpha_2, \ldots, \alpha_n$ be the roots of $f(x)$, and suppose we know the value of

$$s_k = \alpha_1^k + \alpha_2^k + \cdots + \alpha_n^k$$

for some k. Write s_k in the form:

$$s_k = \alpha_1^k \left(1 + \frac{\alpha_2^k}{\alpha_1^k} + \cdots + \frac{\alpha_n^k}{\alpha_1^k}\right).$$

Let α_1 be the root whose absolute value is greatest. Hence, for a sufficiently large k, the numbers α_i^k/α_1^k are negligible, so $s_k \approx \alpha_1^k$, or:

$$\lim_{n\to\infty} s_k^{1/k} = \alpha_1.$$

Consider now transformation 9; applying it several times we obtain a polynomial of degree $n = \partial f$ whose roots are $\alpha_1^{2^m}, \alpha_2^{2^m}, \ldots, \alpha_n^{2^m}$:

$$g(x) = R_y(f(x), x - y^{2^m}).$$

Let:

$$g(x) = a_0^{(m)} x^n + a_1^{(m)} x^{n-1} + \cdots + a_n^{(m)};$$

now,

$$-\frac{a_1^{(m)}}{a_0^{(m)}} = \sum_{i=1}^{n} \alpha_i^{2^m}.$$

If the roots are pairwise distinct and $|\alpha_1| > |\alpha_i|$, $i = 1, 2, \ldots, n$, we have, for a sufficiently large m,

$$-\frac{a_1^{(m)}}{a_0^{(m)}} \approx \alpha_1^{2^m}.$$

Analogously:

$$\frac{a_2^{(m)}}{a_0^{(m)}} = \sum_{i,j} (\alpha_i \alpha_j)^{2^m},$$

and if $|\alpha_2| > |\alpha_i|$, $i = 3, 4, \ldots, n$,

$$\frac{a_2^{(m)}}{a_0^{(m)}} \approx (\alpha_1 \alpha_2)^{2^m}.$$

Dividing by $\alpha_1^{2^m} \approx -a_1^{(m)}/a_0^{(m)}$ yields:

$$\frac{a_2^{(m)}}{a_1^{(m)}} \approx -\alpha_2^{2^m},$$

and so forth. If f has just simple real roots, these formulas hold for a sufficiently large m. The sign of the roots is not given, but can be found by substitution or with other methods. However, if f has real coefficients and complex roots, then these roots are pairwise conjugate so they have the same absolute value in pairs. For instance, all the roots of $f = x^n - 1$ have absolute value 1, and in this case the method does not hold.

Example. Let $f = x^3 - 6x^2 + 11x - 6$ (the roots are 3, 2, and 1). We have:

$$R(f(y), x - y^2) = x^3 - 14x^2 + 49x - 36.$$

Here $m = 1$ and $\sqrt{14} = 3.741657\ldots$, $\sqrt{49/14} = 1.870828\ldots$ and $\sqrt{36/49} = 0.857142\ldots$, which approximate 3, 2 and 1, respectively. A better approximation is obtained in the next step, for $m = 2$:

$$R(f(y), x - y^4) = x^3 - 98x^2 + 1393x - 1296.$$

(Note that the ratio of the last two coefficients is already close to 1.) Now, $98^{1/4} = 3.146346\ldots$, $(1393/98)^{1/4} = 1.941696\ldots$ and $(1296/1393)^{1/4} = 0.982117\ldots$. For $m = 4$ we have already (to six decimal digits): 3.000285; 1.999811; 0.999999. As can be seen in this example, the method converges quickly (the convergence is quadratic). However, the coefficients of the corresponding resultants become soon extremely large (in our example, for $m = 3$ there are coefficients with 13 digits) so in practice using resultants is not very useful for problems like the present one.

10. *Algebraic extensions.* If $F \supseteq K$, and θ is an element of F algebraic over K, that is, a root of a polynomial with coefficients in K, take a polynomial $p(x)$ of least degree m among all those θ is a root of. Firstly, $p(x)$ is irreducible, since if $p(x) = h(x)q(x)$, with $\partial h < m$ and $\partial q < m$, then $0 = p(\theta) = h(\theta)q(\theta)$, so either $h(\theta) = 0$ or $q(\theta) = 0$; in both cases θ would be a root of a polynomial of degree less than the degree of $p(x)$. Moreover, $p(x)$ divides all the polynomials admitting θ as a root. Indeed, if $g(\theta) = 0$, then from $g(x) = p(x)q(x) + r(x)$, $\partial r < m$, we have $0 = g(\theta) = p(\theta)q(\theta) + r(\theta)$ and $r(\theta) = 0$, so, from m being the least possible, we have $r(x) = 0$. Now, up to dividing the coefficients of $p(x)$ by the leading coefficient, we may assume $p(x)$ to be monic. So $p(x)$ is unique, since if $p_1(x)$ is another monic polynomial of which θ is a root, of least degree m, then the difference $p(x) - p_1(x)$ would again admit θ as a root and would be of degree at most $m - 1$. So we may talk about *the* minimal polynomial of θ.

Let $f(x)$, of degree n, be the minimal polynomial of an element θ over a field K. The elements $1, \theta, \theta^2, \ldots, \theta^{n-1}$ form a basis of the vector space $K(\theta)$ of the polynomials in θ of degree $< \partial f$ over K. Let $\alpha \in K(\theta)$; then α is a polynomial in θ of degree at most $n-1$, $\alpha = p(\theta) = c_0 + c_1\theta + \ldots + c_{n-1}\theta^{n-1}$. Like all elements of $K(\theta)$, α is algebraic over K: the $n+1$ elements $1, \alpha, \alpha^2, \ldots, \alpha^n$ are linearly dependent over K; so there is a linear combination of these elements that is equal to 0, and the coefficients of this combination are the coefficients of a polynomial that is zero at α. Thus, let us see how to find a polynomial that is zero at α. Consider the product of the basis $\{\theta^i\}$ by α; this product is a linear transformation of the vector space $K(\theta)$. We have:

$$1 \cdot \alpha = c_0 + c_1\theta + \cdots + c_{n-1}\theta^{n-1},$$
$$\theta \cdot \alpha = c_{0,1} + c_{1,1}\theta + \cdots + c_{n-1,1}\theta^{n-1},$$
$$\vdots$$
$$\theta^{n-1}\alpha = c_{0,n-1} + c_{1,n-1}\theta + \cdots + c_{n-1,n-1}\theta^{n-1},$$

having reduced modulo $f(x)$. Therefore, the following system of homogeneous linear equations:

$$\begin{cases} (c_0 - \alpha)x_0 + c_1x_1 + \cdots + c_{n-1}x_{n-1} & = 0, \\ c_{0,1}x_0 + (c_{1,1} - \alpha)x_1 + \cdots + c_{n-1,1}x_{n-1} & = 0, \\ \cdots\cdots\cdots\cdots\cdots\cdots\cdots\cdots\cdots\cdots\cdots\cdots \\ c_{0,n-1}x_0 + c_{1,n-1}x_1 + \cdots + (c_{n-1,n-1} - \alpha)x_{n-1} = 0, \end{cases}$$

admits the non-zero solution $1, \theta, \ldots, \theta^{n-1}$. So its determinant is zero. This determinant is a polynomial g in α, of degree n with coefficients in K; writing it as a polynomial in x, we have $g(\alpha) = 0$, so this is the polynomial we are looking for. It is the characteristic polynomial of the linear transformation induced by α as described. (It is easy to see that, by changing bases in $K(\theta)$, the polynomial $g(x)$ does not change; so we may talk about the characteristic polynomial of the element α.)

Let us see now how to obtain $g(x)$ as a resultant. Set $\alpha = \alpha_1$, $\theta = \theta_1$, $\alpha_2 = p(\theta_2)$, ..., $\alpha_n = p(\theta_n)$, where θ_i, $i > 1$, are the further roots of $f(x)$ in its splitting field[2]. The elements α_i are roots of the same polynomial $g(x)$, as can be seen by substituting in that system α_i for α and the corresponding θ_i for θ. The polynomial $\prod_{i=1}^{n}(x - p(\theta_i))$ admits the roots α_i, so, up to a constant, is equal to g. From this follows, up to a constant,

$$g(x) = R_y(f(y), x - p(y)).$$

Examples. 1. Let $K = Q$, the rational field, and let $\theta = \sqrt{2} + \sqrt{3}$. The polynomial $f(x) = x^4 - 10x^2 + 1$ admits θ as a root (as it is immediate to see from $\theta^2 = 5 + 2\sqrt{6}$ and $(\theta^2 - 5)^2 = \theta^4 - 10\theta^2 + 25 = (2\sqrt{6})^2 = 24$). This polynomial is irreducible over Q, so it is the minimal polynomial of θ. In terms of θ, $\sqrt{2} = (\theta^3 - 9\theta)/2$ (and $\sqrt{3} = (11\theta - \theta^3)/2$). With $f(y) = y^4 - 10y^2 + 1$ and $p(y) = \frac{1}{2}y^3 - \frac{9}{2}y$, we have, after computations:

$$g(x) = R_y(f(y), x - p(y)) = 16(x^4 - 4x^2 + 4) = (4(x^2 - 2))^2.$$

We have indeed $g(\alpha) = g(\sqrt{2}) = 0$. The fact that $g(x)$ is a power of another polynomial (with an exponent greater than 1) corresponds to the fact that $\alpha = \sqrt{2}$ is not a primitive element for the extension, that is, $Q(\sqrt{2})$ is not the whole of $Q(\theta)$ (since, for instance, it is not possible to write $\sqrt{3}$ as a polynomial in $\sqrt{2}$ with coefficients in Q). The exponent m of the power is the degree of $Q(\theta)$ over $Q(\alpha)$, and the base is the minimal polynomial of α over Q (up to a constant). We have $m = 1$ if and only if α is a primitive element of $Q(\theta)$.

The polynomial $g(x)$ is always of the form $h(x)^m$. Indeed, consider, in general, an extension $K(\theta)$ of $K(\alpha)$ of degree m, and let $e_1, e_2, \ldots, e_m$ be a basis of $K(\theta)$ over $K(\alpha)$. Let next $1, \alpha, \ldots, \alpha^{n-1}$ be a basis of $K(\alpha)$ over K, with α of degree n and with polynomial $h(x) = c_0 + c_1 x + \cdots + c_{n-1}x^{k-1} - x^k$. We have $[K(\theta) : K] = mn$, and the mn products $e_i \alpha^j$, $i = 1, 2, \ldots, m$, $j = 1, 2, \ldots, n$, are a basis of $K(\theta)$ over K. The subspace v_i, having basis $e_i, e_i\alpha, \ldots, e_i\alpha^{k-1}$, is invariant under α:

$$e_i \cdot \alpha = e_i\alpha,$$
$$e_i\alpha \cdot \alpha = e_i\alpha^2,$$
$$\vdots$$
$$e_i\alpha^{k-1} \cdot \alpha = e_i\alpha^k = c_0 e_i + c_1 e_i\alpha + \cdots + c_{n-1}e_i\alpha^{n-1}.$$

---

[2] The existence of this field will be proved in Theorem 4.5.

From this follows that α induces on V_i the linear transformation described by the $n \times n$ matrix

$$T = \begin{pmatrix} 0 & 1 & 0 & 0 & \cdots & \cdots & 0 \\ 0 & 0 & 1 & 0 & \cdots & \cdots & 0 \\ \cdots & & \cdots & & \cdots & & \cdots \\ 0 & 0 & 0 & 0 & \cdots & \cdots & 1 \\ c_0 & c_1 & c_2 & c_3 & \cdots & \cdots & c^{k-1} \end{pmatrix}.$$

The element α induces the same linear transformation T on every V_i, so the linear transformation induced by α has on $K(\theta)$ the matrix $T' = \text{diag}(T, T, \ldots, T)$. The matrix T is the companion matrix of $h(x)$: the characteristic polynomial of T is exactly $h(x)$, and that of T' is $h(x)^m$. In the basis $1, \theta, \ldots, \theta^n$, where $n = mk$, the polynomial of α is $g(x)$ and, since the characteristic polynomial does not change under changes of basis, we have $g(x) = h(x)^m$.

2. If α and β are two algebraic numbers over the same field K, the polynomial admitting as a root $\alpha + \beta$ is $R_x(f(x), g(x-y))$, where $f(x)$ and $g(x)$ are the polynomials of α and β, respectively. So, for instance, for the polynomial admitting as a root $\theta = \sqrt{2} + \sqrt{3}$ we have seen above:

$$f(x) = x^4 - 10x^2 + 1 = R_x(x^2 - 2, (x - y)^2 - 3).$$

Similarly, for $\sqrt{2} + \sqrt{3} + \sqrt{5}$ we have:

$$R_y(R_x(x^2 - 2, (x - y)^2 - 3), (y - z)^2 - 5)(= z^8 - 40z^6 + 352z^4 - 960z^2 + 576).$$

References

[**A**], page 242ff.

4

Factoring polynomials

4.1 Kronecker's method

In this first section of the chapter we consider a classical method, due to Kronecker, for factoring an integer polynomial. We should mention at the outset that the interest of the method does not lie in its applicability (it is not really efficient, even for polynomials of degree as low as five) but rather in its existence. Indeed, it affords an algorithm that in finitely many steps allows one to establish whether a polynomial is reducible or not, and if the answer is in the positive it yields the factorisation into irreducible factors. In other words, it allows one to state that the problem of factoring an integer polynomial is decidable.

Let us begin with Gauss lemma. An integer polynomial is said to be *primitive* if the gcd of its coefficients is 1.

Lemma 4.1 (Gauss). *The product of two primitive polynomials is primitive.*

Proof. Let g and h be primitive, and suppose that their product gh is not. Then there exists a prime number p dividing all the coefficients of gh. Let a_s be the first coefficient of g and b_t that of h not divided by p. Now, the coefficient of x^{s+t} in gh is

$$c_{s+t} = \sum_{i+j=s+t} a_i b_j = (a_0 b_{s+t} + a_1 b_{t+s-1} + \cdots + a_{s-1} b_{t+1})$$
$$+ (a_{s+t} b_0 + a_{s+t-1} b_1 + \cdots + a_{s+1} b_{t-1}) + a_s b_t,$$

and p divides the two quantities in brackets because of the minimality of s and t. By assumption, p divides c_{s+t}; it follows that p divides either a_s or b_t, contrary to the hypothesis. ◇

This lemma allows us to prove the following theorem, also due to Gauss.

Theorem 4.1. *If an integer polynomial splits into the product of two rational polynomials, then it also splits into the product of two integer polynomials.*

Machì A.: Algebra for Symbolic Computation.
DOI 10.1007/978-88-470-2397-0_4, © Springer-Verlag Italia 2012

Proof. Let $f = gh$ with g and h rational. Reduce the coefficients of g and h to the same denominator and collect the gcd of the coefficients of the numerators:

$$f = af_1, \ g = \frac{b}{m}g_1, \ h = \frac{c}{n}h_1.$$

Then $amnf_1 = bcg_1h_1$, where g_1 and h_1 are primitive. The gcd of the coefficients of the left hand side is amn, and of the right side is the bc because by the lemma g_1h_1 is primitive. It follows $a = \frac{bc}{mn}$, so $\frac{bc}{mn}$ is an integer. Hence $f = (ag_1)h_1$, where ag_1 and h_1 are integer polynomials.   ◇

Remark. The original version of Gauss lemma is the following:

> *Let g and h be monic polynomials with rational coefficients. If these coefficients are not all integer, then the coefficients of $f = gh$ cannot all be integer.*

The modern version we have given implies the original one (the converse is also true, but not immediate). However, Gauss version has the advantage of lending itself to a very deep generalisation:

> *Let g and h be monic polynomials whose coefficients are algebraic numbers. If these coefficients are not all algebraic integers, then the coefficients of $f = gh$ cannot all be algebraic integers.*

Our version does not lend itself to a similar generalisation because there is no notion of a gcd for algebraic numbers. However, there exists a formulation due to Dedekind that generalises to algebraic numbers:

> *Let g and h be rational polynomials. If all the coefficients of the product gh are integer, then every coefficient of g times every coefficient of h is an integer.*

The generalisation to algebraic numbers is the "Prague theorem" also due to Dedekind:

> *Let g and h be polynomials whose coefficients are algebraic numbers. If all the coefficients gh are algebraic integers, then every coefficient of g times every coefficient of h is an algebraic integer.*

If we seek a factorisation of an integer polynomial f, then, by the theorem, it is sufficient to seek a factorisation $f = gh$, where the coefficients of g and h are integers. The method by Kronecker we are now going to discuss allows us to determine, if any, such a factorisation, or to prove that the polynomial is irreducible. It works as follows. In a factorisation $f = gh$, with $\partial f = n$, the degree of one of the factors, g say, is at most $\lfloor \frac{n}{2} \rfloor = m$, and therefore g is determined when its values on $m + 1$ distinct points are known. Let $x_0, x_1, \ldots, x_m$ be integers; since $f(x_i) = g(x_i)h(x_i)$, the possible values for $g(x_i)$ must be found among the divisors of the integer $f(x_i)$. We proceed as follows:

1. Evaluate $f(x_i)$, $i = 0, 1, \ldots, m$.

2. For each i find the divisors of $f(x_i)$.

3. For each $(m+1)$-tuple $(d_0, d_1, \ldots, d_m)$ of divisors d_i of $f(x_i)$, set $g(x_i) = d_i$.

4. Each of these choices gives the values of g at the $m + 1$ points x_i.

5. Lagrange's method allows us to uniquely determine a polynomial
$g = g(x) = d_0 L_0(x) + d_1 L_1(x) + \cdots + d_m L_m(x)$.

6. If the coefficients of this polynomial are not integer, or they are integer but this polynomial does not divide f, reject it and try another $(m+1)$-tuple of divisors.

7. $f(x_i)$ being an integer, it admits only finitely many divisors, so there is only a finite number of $(m+1)$-tuples and hence of polynomials g.

8. If no g found in this way divides f, the polynomial f is irreducible over Z, and therefore (Gauss lemma) also over Q.

Hence this method reduces the search for divisors of a polynomial to that of divisors of integers.

Example. Let $f(x) = x^5 + x - 1$. Since one of the two possible factors $g(x)$ has degree at most 2, let us take three points $x_0 = 0$, $x_1 = 1$ and $x_2 = -1$. We have

$$f(0) = -1, \quad g(0) \;= d_0 = 1, -1,$$
$$f(1) = \;\;\; 1, \quad g(1) \;= d_1 = 1, -1,$$
$$f(-1) = -3, \quad g(-1) = d_2 = 1, -1, 3, -3,$$

and the Lagrange's polynomials for the points 0, 1 and -1 are:

$$L_0(x) = -x^2 + 1, \; L_1(x) = \frac{x^2 + x}{2}, \; L_2(x) = \frac{x^2 - x}{2}.$$

The value $d_0 = -1$ can be rejected (just change g into $-g$). The triple $d_0 = 1$, $d_1 = 1$, $d_2 = 1$ yields $g = 1$; there are seven triples remaining. At this point, before going on with other triples of divisors, we may take an $\bar{x}$ different from all the x_i, and see for what triples d_0, d_1, d_2 the number

$$s = d_0 L_0(\bar{x}) + d_1 L_1(\bar{x}) + d_2 L_2(\bar{x})$$

(which gives the value of $g(\bar{x})$) is an integer and divides $f(\bar{x})$. If this is not the case, the triple is rejected. In our case, let $\bar{x} = 2$; the above becomes

$$s = -3d_0 + 3d_1 + d_2.$$

Hence we have to consider only the triples for which s is an integer that divides $f(2) = 33$. For instance, the triple $1, -1, 1$ yields $s = -5$, and therefore is rejected. The triple $1,1,3$ gives $s = 3$. With these values of the d_is we find $g = x^2 - x + 1$, which indeed divides f:

$$x^5 + x - 1 = (x^2 - x + 1)(x^3 - x^2 + 1).$$

In order to see if these factors are irreducible we continue the procedure. For example, with the triple $1, -1, 3$ have $s = -3$ and $g = -2x + 1$, which does not divide f because f does not have the root $\frac{1}{2}$. With $1, 1, -3$ we find

$g = -x^2 + x + 2$, which is not a factor of f because its constant term does not divide that of f. The remaining triples are discarded in a similar way, and the one we have found is the factorisation into irreducible factors.

Besides considering the *control value* $\overline{x}$ as done in the example, there are other means that allow us to simplify calculations:

1. Evaluate $f(x_i)$ at more than $m + 1$ points x_i so that we may choose the $f(x_i)$s that have the least number of divisors.
2. For an integer q, the polynomial $g(x) - g(q)$ has the root q, and therefore is divisible by $x - q$, and the quotient has integer coefficients. In particular, for $x = p$, $g(p) - g(q)$ is divisible by $p - q$. It follows that $g(x_i) - g(x_k)$ is divisible by $x_i - x_k$, i.e. $x_i - x_k$ divides $d_i - d_k$. In this way, it is possible to discard the group of divisors d_i for which this condition is not satisfied[1].

As we have said, this method is not efficient. Modern, more efficient methods consist in factoring the polynomial modulo a prime number p and then lift this factorisation to one over the integers. First, let us see if it is possible to find out whether the polynomial is irreducible.

4.2 Irreducibility criteria

If f is a primitive polynomial, taking the coefficients modulo a prime p we obtain a polynomial $\overline{f}$ over Z_p whose degree is the same as that of f if p does not divide the leading coefficient of f. If f reduces over Z, $f = gh$ with g and h primitive, then $\overline{f} = \overline{g}\overline{h}$. Hence, if for some p we find that $\overline{f}$ is irreducible, then f is irreducible over the integers. This gives a condition for irreducibility.

Example. Let $f = x^2 + x + 1$. If f reduces modulo some prime p then it reduces into two polynomial of first degree, and hence has two roots in Z_p. But, for $p = 2$, the polynomial has no roots, hence is irreducible over Z_2 and so also over Z.

One may then try other primes, using also the fact that Z_p is a field, and therefore the factorisation into irreducible factors is unique.

Example. Let $f = x^4 + 3x^3 + 3x^2 - 5$. Modulo 2, this polynomial is $x^4 + x^3 + x^2 + 1$, and admits the root 1. Dividing by $x - 1$ we find

$$f = (x + 1)(x^3 + x + 1);$$

the two factors being irreducible, the latter is *the* factorisation of f over Z_2. Hence, if f factors over Z, this cannot happen with an irreducible factor of degree 2. Indeed, either such a factor remains irreducible over Z_2, but this would yield a different factorisation over Z_2, or it decomposes into two linear

---

[1] This remark is due to Runge.

factors, which would give yet another factorisation mod 2. The only possibility is that, over Z, $f = (ax + b)q(x)$, so that f has a linear factor, and so a root, modulo p for all p. But for $p = 3$, $f = x^4 + 1$, and $f(0) = 1$, $f(1) = 2$ and $f(2) = 2$, so there are no roots over Z_3. It follows that f is irreducible over Z.

A sufficient condition for the irreducibility of a polynomial is given in the following theorem.

Theorem 4.2 (Eisenstein's criterion). *Let*

$$f = a_n x^n + a_{n-1} x^{n-1} + \cdots + a_1 x + a_0 \tag{4.1}$$

be an integer polynomial. If there exists a prime p such that $p \nmid a_n$, $p | a_i$ for $i < n$, and $p^2 \nmid a_0$, then f is irreducible over Z.

Proof. Let $f = gh$ over Z. Taking the coefficients mod p have $\bar{g}\bar{h} \equiv \bar{a}_n x^n$. The factorisation over Z_p being unique, the irreducible factors of $\bar{g}$ must be among those of $\bar{a}_n x^n$, and therefore their constant term is zero. The same holds for $\bar{h}$. It follows that if b_0, c_0 and a_0 are the constant terms of g, h and f, respectively, we have $p | b_0$, $p | c_0$, from which $p^2 | b_0 c_0 = a_0$, against the assumption. ◇

This criterion can either be applied directly, for instance in the case of the polynomial $f = x^4 + 3x^3 + 6x^2 - 15$, with $p = 3$, or after a suitable transformation of the polynomial, as the following example shows.

Example. Let $f = x^4 + 1$. As it stands, the polynomial does not have the form required for the criterion to be applicable. But since if $f(x)$ is reducible so is $f(x + 1)$, let us replace x with $x + 1$; we obtain

$$f(x + 1) = (x + 1)^4 + 1 = x^4 + 4x^3 + 6x^2 + 4x + 2.$$

The criterion is now applicable with $p = 2$, and we conclude that f is irreducible over Z.

Lemma 4.2. *If p is a prime, and $0 < k < p$, then the binomial coefficient $\binom{p}{k}$ is divisible by p.*

Proof. We have

$$\binom{p}{k} = \frac{p!}{k!(p-k)!},$$

and if $k \neq p$, $k \neq 0$ the denominator is not divisible by p. ◇

Example. Let, with p a prime,

$$f = \frac{x^p - 1}{x - 1} = x^{p-1} + x^{p-2} + \cdots + x + 1.$$

We replace x with $x + 1$:

$$\frac{(x + 1)^p - 1}{(x + 1) - 1} = \frac{1}{x}\left(\sum_{k=0}^{p} \binom{p}{k} x^k - 1\right).$$

Now, for $0 < k < p$ the binomial coefficient $\binom{p}{k} = \frac{p!}{k!(p-k)!}$ is divisible by p. Hence,

$$f(x+1) = \frac{1}{x}(1 + px + \cdots + x^p - 1) = x^{p-1} + \cdots + p,$$

where the dots denote terms divisible by p. Now the criterion is applicable, and we can conclude that this polynomial f is irreducible over Z.

Remark. The polynomials of the previous two examples are *cyclotomic* polynomials (polynomials of the division of the circle). They are $\varphi_8(x)$ and $\varphi_p(x)$, respectively. We will come back to this later.

Eisenstein's criterion admits the following generalisation (that we state without proof): let f be the polynomial (4.1); if there exist a prime p and an integer i such that $\gcd(i, n) = 1$ and $p^i \nmid a_n$, $p^i | a_k$ for $k < n$, and $p^{i+1} \nmid a_0$, then f is irreducible.

Example. Let $f = x^5 + 4x^3 + 8x^2 + 8x + 4$. With $p = 2$ Eisenstein's criterion is not applicable, but the generalised one is applicable with $i = 2$. Note that the condition that i be coprime with the degree is necessary. The above polynomial with x^5 replaced by x^4 reduces as $(x^2 + 2x + 2)^2$.

We may conclude as follows. Before applying Kronecker's method, perhaps only to discover that the polynomial is irreducible, one should try to know beforehand that the polynomial is irreducible, by applying the above criteria or considering the polynomial modulo one or more primes. However, the search of a prime for which the polynomial is irreducible may be vain. Indeed, there exist polynomials that split modulo every prime number but are irreducible over the integers (we will see some examples in the next section). If these or other attempts fail, Kronecker method may be applied. This was more or less the situation in 1960s. Nowadays much more effective methods are available, as we shall see in a moment. We have to review, beforehand, a few properties of the finite fields and of the polynomials over these fields.

4.3 Finite fields and polynomials

Let Z_p be the field of congruence classes mod p, $Z_p = \{0, 1, \ldots, p-1\}$. We will write the latter elements as

$$-\frac{p-1}{2}, -\frac{p-3}{2}, \ldots, -1, 0, 1, \ldots, \frac{p-3}{2}, \frac{p-1}{2},$$

the so-called *balanced representation*. In any field, an equality $a^2 = b^2$ is only possible if $a = \pm b$, as it follows from the equality $0 = a^2 - b^2 = (a-b)(a+b)$ and the absence of zero divisors. Let Z_p^* denote the set (the group) of non-zero elements of Z_p. By what we have just seen, the squares of the elements $0, 1, \ldots, \frac{p-1}{2}$ are all distinct, and are equal to those obtained by squaring the remaining elements. The next two corollaries follow from this fact.

Corollary 4.1. *Let $p > 2$. Then exactly half of the elements of Z_p^* are squares.*

Example. The squares of Z_{13}^* are $1, 4, -4, 3, -1, -3$.

It is clear that the product of two squares is again a square. If $Q = \{a_1, a_2, \ldots, a_r\}$ are the squares of Z_p^*, and b is not a square, then the set $Qb = \{a_1 b, a_2 b, \ldots, a_r b\}$ consists of distinct elements and is disjoint from Q because if $a_i b = a_k$ then $b = a_i^{-1} a_k$ would be a square. It follows $Z_p^* = Q \cup Qb$. If c and d are non-squares, then $c = a_i b$ and $d = a_k b$, from which $cd = a_i b a_k b = a_i a_k b^2$, the product of two squares, and cd is a square.

Corollary 4.2. *In Z_p^* the product of two squares is a square, and the product of two non-squares is a square.*

Corollary 4.3. *Let -1, a and $-a$ be three elements of Z_p^*. Then one of the three is a square.*

Proof. If -1 and a are non-squares, by the previous corollary their product $-1 \cdot a = -a$ is a square. $\Diamond$

Corollary 4.3 is especially useful to prove that there exist irreducible polynomials over the integers that are reducible modulo every prime.

Examples. 1. Using Eisenstein's criterion we have seen that $x^4 + 1$ is irreducible over Z. Let us now prove that it splits over Z_p for every prime p:

i)   if -1 is a square mod p, $-1 = a^2$, then

$$x^4 + 1 = x^4 - (-1) = x^4 - a^2 = (x^2 - a)(x^2 + a);$$

ii)   if 2 is a square mod p, $2 = a^2$, then

$$x^4 + 1 = (x^2 + 1)^2 - 2x^2 = (x^2 + 1)^2 - (a^2 x^2) = (x^2 + 1)^2 - (ax)^2$$
$$= (x^2 + 1 - ax)(x^2 + 1 + ax);$$

iii)   if -2 is a square mod p, $-2 = a^2$, then

$$x^4 + 1 = (x^2 - 1)^2 + 2x^2 = (x^2 - 1)^2 - (-2x^2) = (x^2 - 1)^2 - (ax)^2$$
$$= (x^2 - 1 - ax)(x^2 - 1 + ax).$$

(As one can see, the technique is that of *completing the square*.) So, for $p = 5$, -1 is a square: $-1 = 4 = 2^2$, and we are in case *i)*:

$$x^4 + 1 = (x^2 - 2)(x^2 + 2).$$

For $p = 23$, $2 = 5^2$, and from *ii)* we have

$$(x^2 - 5x + 1)(x^2 + 5x + 1).$$

For $p = 11$, $-2 = 3^2$, and from iii) we have

$$x^4 + 1 = (x^2 - 3x - 1)(x^2 + 3x - 1).$$

In case ii), the discriminant of the two factors is $a^2 - 4$, i.e. -2. Hence, if -2 is also a square, $-2 = b^2$, the two factors have the roots $(a \pm b)/2$ and $(-a \pm b)/2$ (here dividing by 2 means multiplying by the inverse of two, which always exists if $p > 2$). It follows

$$x^4 + 1 = \left(x - \frac{a+b}{2}\right)\left(x - \frac{a-b}{2}\right)\left(x - \frac{-a+b}{2}\right)\left(x - \frac{-a-b}{2}\right),$$

and the polynomial splits into linear factors. This happens for instance for $p = 17$, where $2 = 6^2$ and $-2 = 7^2$; moreover, the inverse of 2 is 9. We have:

$$x^4 + 1 = (x - (6+7) \cdot 9)(x - (-1) \cdot 9)(x - 1 \cdot 9)(x - (-6-7) \cdot 9)$$
$$= (x - 117)(x + 9)(x - 9)(x + 117)$$
$$= (x - 2)(x + 9)(x - 9)(x + 2).$$

2. Another polynomial irreducible over Z which splits for all p is $x^4 - 2x^2 + 9$:

i)   if $-1 = a^2$,

$$x^4 - 2x^2 + 9 = (x^2 - 3)^2 + 4x^2 = (x^2 - 3)^2 - (-4x^2) = (x^2 - 3)^2 - (2ax)^2$$
$$= (x^2 - 3 - 2ax)(x^2 - 3 + 2ax).$$

With $p = 5$ we have $-1 = 4 = 2^2$, $a = 2$:

$$x^4 - 2x^2 + 9 = (x^2 - 3 - 4x)(x^2 - 3 + 4x);$$

ii)   if $2 = a^2$,

$$x^4 - 2x^2 + 9 = (x^2 + 3)^2 - 8x^2 = (x^2 + 3)^2 - 2 \cdot 4x^2$$
$$= (x^2 + 3)^2 - (a \cdot 2 \cdot x)^2 = (x^2 + 3 - 2ax)(x^2 + 3 + 2ax).$$

With $p = 17$, $2 = 6^2$, $a = 6$ and

$$x^4 - 2x^2 + 9 = (x^2 + 3 - 12x)(x^2 + 3 + 12x);$$

iii)   if $-2 = a^2$,

$$x^4 - 2x^2 + 9 = (x^2 - 1)^2 + 8 = (x^2 - 1)^2 - (-2 \cdot 4) = (x^2 - 1)^2 - (2a)^2$$
$$= (x^2 - 1 - 2a)(x^2 - 1 + 2a).$$

With $p = 11$, $-2 = 3^2$, $a = 3$ and

$$x^4 - 2x^2 + 9 = (x^2 - 7)(x^2 + 5).$$

In case i) the discriminant of the two factors is $a^2 + 3 = -1 + 3 = 2$. Hence, if 2 is also a square, $2 = b^2$, the polynomial splits into linear factors:

$$x^4 - 2x^2 + 9 = (x + a - b)(x + a + b)(x - a - b)(x - a + b).$$

This happens for instance for $p = 17$, where $-1 = 4^2$ and $2 = 6^2$. Then with $a = 4$ and $b = 6$ we have

$$x^4 - 2x^2 + 9 = (x - 2)(x + 10)(x - 10)(x + 2).$$

A similar discussion may be done in the other cases.

The polynomial seen in the last example has the complex roots $\pm i \pm \sqrt{2}$. More generally (a result we state without proof), if $p_1, p_2, \ldots, p_{n-1}$ are the first $n - 1$ prime numbers, the polynomial whose roots are:

$$\pm i \pm \sqrt{2} \pm \sqrt{3} \pm \cdots \pm \sqrt{p_{n-1}}$$

is of degree 2^n, is irreducible over Z but splits for every p. Such is, for instance,

$$x^8 - 16x^6 + 88x^4 + 192x^2 + 144,$$

whose roots are $\pm i \pm \sqrt{2} \pm \sqrt{3}$.

3. Both $x^4 + 1$ and $x^4 - 2x^2 + 9$ are special cases of

$$f(x) = x^4 + ax^2 + b^2.$$

Let us prove that, whatever a and b, this polynomial factorises for every p. If $p = 2$, then $a, b = 0, 1$, and the polynomial splits: $(x^2 + ax + b)^2$. If $p > 2$, then there exists $s = \frac{a}{2}$ and $f(x) = x^4 + 2sx + b^2$ may be written in three ways:

$$(x^2 + s)^2 - (s^2 - b^2),$$
$$(x^2 + b)^2 - (2b - 2s)x^2,$$
$$(x^2 - b)^2 - (-2b - 2s)x^2.$$

If neither $2b - 2s$ nor $-2b - 2s$ are squares, so is their product, and it equals $4(s^2 - b^2) = c^2$, from which $s^2 - b^2 = \frac{c^2}{4} = (\frac{c}{2})^2$. Hence, in every case, $f(x)$ is a difference of two squares.

Lemma 4.3. *Let a and b be two integers. Then:*

$$(a + b)^p \equiv a^p + b^p \bmod p \tag{4.2}$$

and, for every n,

$$(a + b)^{p^n} \equiv a^{p^n} + b^{p^n} \bmod p. \tag{4.3}$$

More generally:

$$(a_1 + a_2 + \cdots + a_k)^{p^n} \equiv a_1^{p^n} + a_2^{p^n} + \cdots + a_k^{p^n} \bmod p.$$

Proof.

$$(a + b)^p = \sum_{k=0}^{p} \binom{p}{k} a^k b^{p-k}. \tag{4.4}$$

For $k \neq 0, p$, the binomial coefficient $\binom{p}{k}$ is divisible by p (Lemma 4.4), so it is 0 mod p. For $k = 0$ we have b^p, and for $k = p$ we have a^p. If $n = 1$, (4.3) is simply (4.2). If $n > 1$,

$$(a + b)^{p^n} \equiv ((a + b)^{p^{n-1}})^p \equiv (a^{p^{n-1}} + b^{p^{n-1}})^p \equiv a^{p^n} + b^{p^n} \text{ mod } p,$$

where the second equivalence follows by induction and the third one by (4.2). The last claim is obtained by induction.   ◇

Remark. This lemma holds, with the same proof, for any commutative ring of characteristic p.

Theorem 4.3 (Fermat's little theorem). *For every integer a and every prime p,*

$$a^p \equiv a \text{ mod } p, \tag{4.5}$$

and more generally:

$$a^{p^n} \equiv a \text{ mod } p. \tag{4.6}$$

Proof. If $a = 0$ there is nothing to prove. We distinguish two cases, $a > 0$ and $a < 0$.

i) $a > 0$. By induction on a. From $a = (a - 1) + 1$ we have, by the previous lemma,

$$a^p = (a - 1)^p + 1^p.$$

By induction, $(a - 1)^p \equiv a - 1 \text{ mod } p$, so that

$$a^p \equiv a - 1 + 1 = a \text{ mod } p;$$

ii)   if $a < 0$, then $-a > 0$, so by what we have just seen $(-a)^p \equiv -a \text{ mod } p$. If $p = 2$, then $-a \equiv a \text{ mod } 2$; if $p > 2$, p is odd and we have $(-a)^p = -(a^p)$, so $-(a^p) = (-a)^p \equiv -a \text{ mod } p$, where the last congruence follows from the previous case. (4.6) is obtained by induction from (4.5).   ◇

Corollary 4.4. *If $a \not\equiv 0 \text{ mod } p$, then*

$$a^{p-1} \equiv 1 \text{ mod } p.$$

Proof. If $a \not\equiv 0 \text{ mod } p$, a admits an inverse mod p. By multiplying the two sides of (4.5) by this inverse we have the result.   ◇

Corollary 4.5. *We have*

$$x^p - x \equiv (x - 0)(x - 1) \cdots (x - (p - 1)) \bmod p,$$

and more generally, for every integer polynomial $f(x)$,

$$f(x)^p - f(x) \equiv (f(x) - 0)(f(x) - 1) \cdots (f(x) - (p - 1)) \bmod p.$$

Proof. By Fermat's theorem, any integer mod p is a root of $x^p - x$, so this polynomial is divisible by $x - i$, $i \in Z_p$. The other congruence follows from the first one. $\diamondsuit$

Theorem 4.4. *Let K be a field, and let f be an irreducible polynomial with coefficients in K. Then the polynomials over K of degree less than the degree of f form a field F with respect to the usual sum and to the product modulo f.*

Proof. F is closed w.r.t. the stated operations. As to the existence of the inverse of a polynomial $g \in F$, note that $\gcd(f, g) = 1$ since f is irreducible and $\partial g < \partial f$, and therefore $af + bg = 1$, for some polynomials a and b. Modulo f this equality becomes $bg \equiv 1 \bmod f$, so g has the inverse $b \bmod f$. Hence F is a field. Note that F contains K as the set of polynomials of degree zero (the constants). $\diamondsuit$

Corollary 4.6. *Let K and f be as in the previous theorem. Then there exists an extension F of K that contains a root of f.*

Proof. Let F be as in the theorem. Since $f(x) \equiv 0 \bmod f(x)$, the polynomial $p(x) = x$ is a root of f. $\diamondsuit$

Theorem 4.5. *If $f(x)$ is a polynomial with coefficients in the field K, there exists an extension of K in which $f(x)$ splits into linear factors.*

Proof. If $\partial f = 1$, then the required extension is K itself. Otherwise, let $f(x) = g(x)q(x)$ with $g(x)$ irreducible. By the previous theorem there exists $F \supseteq K$ in which g has a root, α say, so that $g(x) = (x - \alpha)h(x)$ over F. Hence $f(x) = (x - \alpha)h(x)q(x)$ over F, and the degree of $h(x)q(x)$ being equal to $\partial f - 1$, by induction on the degree of f there exists an extension of F over which $h(x)q(x)$, and so f, splits into linear factors. $\diamondsuit$

The smallest field K' containing K in which f splits into linear factors is called the *splitting field* of f. It is the "smallest" in the sense that f does not split into linear factors over any subfield of K'.

Theorem 4.6. *The set of roots of the polynomial $x^{p^n} - x$ over Z_p is a field.*

Proof. First, 0 and 1 are roots, as well as all the other elements of Z_p (Theorem 4.3). Moreover, if α and β are roots in an extension of Z_p, so are $\alpha + \beta$ and $\alpha\beta$. Indeed,

$$\begin{aligned}
(\alpha + \beta)^{p^n} - (\alpha + \beta) &= \alpha^{p^n} + \beta^{p^n} - \alpha - \beta \\
&= (\alpha^{p^n} - \alpha) + (\beta^{p^n} - \beta) = 0 + 0 = 0
\end{aligned}$$

(Lemma 4.3), and

$$(\alpha\beta)^{p^n} - \alpha\beta = \alpha^{p^n}\beta^{p^n} - \alpha\beta = \alpha\beta - \alpha\beta = 0.$$

As to the inverse,

$$(\alpha^{-1})^{p^n} - \alpha^{-1} = (\alpha^{p^n})^{-1} - \alpha^{-1} = \alpha^{-1} - \alpha^{-1} = 0,$$

and for the opposite,

$$(-\alpha)^{p^n} - (-\alpha) = (-1)^{p^n}\alpha^{p^n} + \alpha = (-1)^{p^n}\alpha + \alpha;$$

if p is odd the last quantity is $-\alpha+\alpha = 0$, if $p = 2$ we have $\alpha+\alpha = 2\alpha = 0$. $\diamond$

It follows that the set of roots makes up the splitting field of the polynomial $x^{p^n} - x$. (In Corollary 4.5 we have seen that Z_p is the splitting field of $x^p - x$.) Hence this field has at most p^n elements; we will see below (Corollary 4.7) that it has exactly p^n elements.

We recall that a finite field has a number of elements which is a power of a prime p. Indeed, such a field contains Z_p, as the set of multiples of the identity; moreover, it is a vector space over Z_p of finite dimension, m say. Hence its elements are all the linear combinations of the elements of a basis with coefficients in Z_p, and these are p^m in number. If $q = p^n$, we will denote by F_q a field with q elements (we shall see in Corollary 4.9 that it is unique up to isomorphisms). However, we will keep the notation Z_p for the field of residue classes mod p.

Let F_q be a finite field. Then there exists an element $a \in F_q^* = F_q \setminus \{0\}$ such that any other element of F_q^* is a power of a (in other words, the multiplicative group of a finite field is cyclic). Such an element is called a *primitive element* of F_q. We now prove the existence of such an element together with other properties of finite fields.

1. *Existence of a primitive element.* Let $a \in F_q^*$; not all the powers of a can be distinct because F_q is finite. If $a^h = a^k$ with $h > k$, then $a^{h-k} = 1$, and there is a smallest m for which $a^m = 1$; m is the *order* $o(a)$ of a. If $a^k = 1$ for some k, then $o(a)$ divides k.

Let $o(a) = h$ and $o(b) = k$, with $\gcd(h, k) = 1$; then $o(ab) = hk$. First, by commutativity, $(ab)^{hk} = 1$. If $o(ab) < hk$ then $o(ab)|hk$ and therefore $o(ab) = h'k'$, with $h'|h$ and $k'|k$, and so $\gcd(h', k') = 1$. We have $(ab)^{h'k} = 1$, so $a^{h'k}b^{h'k} = 1$, from which $a^{h'k} = 1$. Then $o(a) = h|h'k$, and since $\gcd(h, k) = 1$ we must have $h|h'$, and hence $h' = h$. A similar argument applied to hk' yields the result.

Let a be an element of maximal order M. We show that any other $b \in F_q^*$ has order dividing M. Suppose this is not the case; then there exists a prime r such that $o(b) = r^h u$ and $M = r^k v$, where $\gcd(r, u) = \gcd(r, v) = 1$ and $h > k$. Then $o(a^{r^k}) = v$ and $o(b^u) = r^h$, and since $\gcd(r, v) = 1$, the element $a^{r^k}b^u$ has order the product vr^h of the orders, which is greater than M.

The M powers $1, a, a^2, \ldots, a^{M-1}$ of a are all distinct, and are all roots of $x^M - 1$, which cannot have other roots, since it is of degree M. But every $b \in F_q^*$ is a root of this polynomial because $o(b)|M$ by what we have just seen, and therefore $b^M = 1$, i.e. $b^M - 1 = 0$. It follows that b is one of the powers of a.

If $a \in F_q^*$ has order m, then a^k also has order m if and only if $\gcd(k, m) = 1$. If a^k has order m, and $d = \gcd(k, m) > 1$, then $(a^k)^{\frac{m}{d}} = (a^{\frac{k}{d}})^m = 1$, against the minimality of m. Conversely, if $\gcd(k, m) = 1$ and $(a^k)^h = a^{kh} = 1$, then $m|kh$, and since $\gcd(k, m) = 1$ we have $m|h$. However, m being the maximal possible order for a power of a, we have $h = m$.

It follows that the number of elements of F_q^* of maximal order is $\varphi(q-1)$, where φ is Euler's function.

2. The number of polynomials of a given degree k over F_q is $(q-1)q^k$ (there are $q-1$ choices for the leading coefficient, which must be non zero otherwise the degree is less than k, and q choices for each of the other k coefficients). Hence the number of polynomials of degree less than m (counting also the null polynomial) is

$$1 + \sum_{k=0}^{m-1}(q-1)q^k = 1 + (q-1)\sum_{k=0}^{m-1}q^k = 1 + (q-1)(1 + q + q^2 + \cdots + q^{m-1})$$

$$= 1 + (q-1)\frac{q^m - 1}{q - 1} = 1 + q^m - 1 = q^m.$$

3. *The field F_q, $q = p^n$, contains one and only one subfield of degree p^d for each divisor d of n.* Indeed, let k be such a subfield, with $|K| = p^d$; let us prove that $d|n$. Every element of K is a root of the polynomial $x^{p^d} - x$, but as an element of F_q it is also a root of $x^{p^n} - x$. It follows that $x^{p^d} - x$ divides $x^{p^n} - x$; by Exercise 11, $d|n$. Conversely, if $d|n$ then $x^{p^d} - x$ divides $x^{p^n} - x = \prod_{\alpha \in F_q}(x - \alpha)$, and therefore $x^{p^d} - x$ has p^d roots in F_q. By the argument given for Theorem 4.6, these form a field. The subfield of order p^d consists of the elements of F_q that are roots of $x^{p^d} - x$.

4. *If K_1 and K_2 are two subfields of F_q of order p^r and p^s, respectively, then $|K_1 \cap K_2| = p^{\gcd(r,s)}$.* Indeed, by what we have just seen, K_1 and K_2 both contain the subfield K' of order $p^{\gcd(r,s)}$ because $\gcd(r, s)$ divides r and s. Conversely, every subfield common to K_1 and K_2 has order dividing r and s, and so also $\gcd(r, s)$, and is therefore contained in K'.

Theorem 4.7. *i) The mapping $\phi : F_q \longrightarrow F_q$ given by $x \to x^p$ is an automorphism (Frobenius automorphism) of order m, where $q = p^m$.*

ii) The fixed points of ϕ are all and only the elements of $Z_p \subseteq F_q$.

Proof. i) In any field, $(xy)^p = x^p y^p$. Moreover, by Lemma 4.3, $(x + y)^p = x^p + y^p$, so that ϕ is a homomorphism of fields, whose kernel is 0 because

$x^p = 0$ implies $x = 0$. Hence ϕ is injective and, F_q being finite, also surjective. Moreover, $\phi^m(x) = x^{p^m} = x^q = x$, so that $\phi^m = id$, and it cannot be $\phi^t = id$ with $t < m$ because otherwise $x^{p^t} = x$ for a primitive element x, and therefore $x^{p^t-1} = 1$ with $p^t - 1 < q - 1$.

ii) A fixed point a is such that $a^p = a$, and therefore is a root of $x^p - x$. The p elements of Z_p are roots of this polynomial; being of degree p, it cannot have more than p roots. $\qquad\qquad\qquad\diamond$

Theorem 4.8. *Every polynomial in x^p over F_q is the p-th power of a polynomial in x:*

$$f(x^p) = g(x)^p,$$

for some $g(x)$. Moreover, if $F_q = Z_p$ then $g(x) = f(x)$.

Proof. Let

$$f(x^p) = a_0 + a_1 x^p + a_2 (x^p)^2 + \cdots + a_t (x^p)^t.$$

By Theorem 4.7 i), for each i we have $a_i = b_i^p$, for some b_i, and therefore by Lemma 4.3,

$$f(x^p) = b_0^p + (b_1 x)^p + (b_2 x^2)^p + \cdots + (b_t x^t)^p$$
$$= (b_0 + b_1 x + b_2 x^2 + \cdots + b_t x^t)^p,$$

and with $g(x) = b_0 + b_1 x + b_2 x^2 + \cdots + b_t x^t$ we have the result. If $F_q = Z_p$ then, by Fermat's theorem, $b_i = b_i^p$, from which $b_i = a_i$ and $g(x) = f(x)$. $\quad\diamond$

Theorem 4.9. *If the derivative $f'(x)$ of $f(x)$ is zero over F_q, then $f(x)$ is a polynomial in x^p:*

$$f(x) = h(x^p) = g(x)^p, \qquad\qquad (4.7)$$

where the last equality follows from the previous theorem. Conversely, if (4.7) holds, then $f'(x) = 0$.

Proof. Let $f(x) = \sum_{k=0}^n a_k x^k$; then $f'(x) = \sum_{k=1}^n k a_k x^{k-1}$, and if this polynomial is the null polynomial, then p divides all the coefficients $k a_k$, and therefore either $p | a_k$ or $p | k$. If $p | a_k$, then $a_k \equiv 0 \bmod p$, and therefore the monomial $a_k x^k$ is not present in $f(x)$. Hence the monomials $a_k x^k$ that are present are at most those for which $p | k$, i.e. those for which x^k has the form $x^{pt} = (x^p)^t$. Then $f(x)$ is a polynomial in x^p: $f(x) = a_0 + a_p x^p + a_{2p}(x^p)^2 + \cdots + a_{tp}(x^p)^t = h(x^p)$, where some of the a_{kp} may be zero. It follows, by Theorem 4.8, that $f(x) = g(x)^p$. Conversely, if this equality holds, then $f'(x) = p g(x)^{p-1} g'(x) = 0$. $\quad\diamond$

Remark. $f(x)$ is a polynomial in x of degree pt; setting $y = x^p$, we have $h(y) = b_0 + b_1 y + \cdots + b_t y^t$, with $b_k = a_{kp}$, so that $h(y)$ is of degree t.

Theorem 4.10. *Let $f(x)$ be a polynomial over a field K and let $p(x)$ be a factor of $f(x)$ of multiplicity at least $t > 1$. Then $p(x)$ is a factor of the derivative $f'(x)$ of multiplicity at least $t - 1$. In particular, $f(x)$ has no multiple factors if and only if $\gcd(f(x), f'(x)) = 1$.*

Proof. Let $f(x) = p(x)^t q(x)$, $\partial p(x) \geq 1$, $t > 1$. Then

$$f'(x) = tp(x)^{t-1}p'(x)q(x) + p(x)^t q'(x)$$
$$= p(x)^{t-1}(tp'(x)q(x) + p(x)q'(x)),$$

and therefore $p^{t-1}(x)$ is a polynomial of degree at least 1 that divides both $f(x)$ and $f'(x)$; in this case, $\gcd(f(x), f'(x)) \neq 1$.

Conversely, let $p(x)$ be a non constant irreducible polynomial that divides both $f(x)$ and $f'(x)$. Then $f(x) = p(x)q(x)$ from which $f'(x) = p(x)q'(x) + p'(x)q(x)$; since $p(x)$ divides $f'(x)$ it must also divide the product $p'(x)q(x)$, and being irreducible also one of the two factors. If $q(x) = p(x)q_1(x)$, we have $f(x) = p^2(x)q_1(x)$, and $f(x)$ has a multiple factor. If $p(x)$ divides $p'(x)$, then $p'(x) = 0$, the null polynomial, because were it not null it would have a degree $d < \partial p(x)$, and therefore it could not be $p(x)|p'(x)$. (In characteristic zero this case cannot take place because $p(x) = ax^m + \cdots$ with $m > 0$, so that $p'(x) = max^{m-1} + \cdots \neq 0$.) If the characteristic is p, it is possible that $p'(x) = 0$, but then, by the previous theorem, $p(x)$ has the form $g(x)^p$, for some $g(x)$, and therefore cannot be irreducible. $\diamond$

A polynomial that has no multiple factors is said to be *square-free*.

Corollary 4.7. *The splitting field of $x^{p^n} - x$ over Z_p has exactly p^n elements.*

Proof. The polynomial is coprime with its derivative $p^n x^{p^n - 1} - 1 = -1$, and therefore splits into distinct linear factors. $\diamond$

Corollary 4.8. *For every p and n, p a prime, there exists a field with p^n elements.*

Theorem 4.11. *For every p and n, p a prime, there exist irreducible polynomials over Z_p of degree n.*

Proof. Let a be a primitive element of the splitting field F of $x^{p^n} - x$. Since a is a root of $x^{p^n - 1} - 1$, the minimal polynomial $p(x)$ of a over Z_p is of degree $m \leq p^n - 1$. We know that the polynomials of degree less than m over Z_p form a field F_1 with p^m elements. Consider the mapping $\psi : F_1 \to F$ that associates the element $f(a)$ with $f(x) \in F_1$ ($f(a)$ is a linear combination with coefficients in Z_p of powers of a, and so it belongs to F). It is clear that it is a homomorphism. Since $\partial f < m$, if $f(a) = 0$ then f is the null polynomial, and ψ is injective. If $b \in F$, then $b = a^k$ for some k. If $k < m$, then b is the image of the polynomial x^k of F_1. If $k \geq m$, by dividing x^k by $p(x)$ we obtain $x^k = p(x)q(x) + r(x)$, where $\partial r(x) < \partial p(x)$, and therefore $r(x) \in F_1$. Since $p(a) = 0$, we have $a^k = r(a)$, and b is the image of $r(x)$. Hence ψ is also surjective, and therefore an isomorphism. It follows that $p^m = |F_1| = |F| = p^n$, and $m = n$. The polynomial $p(x)$ is irreducible and has degree n. $\diamond$

Corollary 4.9. *Two finite fields of the same order are isomorphic.*

Proof. Let F be finite, $|F| = p^n$. If a is a primitive element of F, then with the same proof as for Theorem 4.11 the minimal polynomial of a has degree n and F is isomorphic to the field of polynomials over Z_p of degree less than n. $\diamond$

Theorem 4.12. *The polynomial $x^{p^n} - x$ over Z_p is the product of all the monic irreducible polynomials $f_d(x)$ over Z_p of degree d that divides n:*

$$x^{p^n} - x = \prod_{d|n} f_d(x). \tag{4.8}$$

Proof. Let $d|n$, $n = dk$. The splitting field of f_d has p^d elements so that if α is a root of f_d then $\alpha^{p^d-1} = 1$, or $\alpha^{p^d} = \alpha$. But

$$\alpha^{p^n} = \alpha^{p^{dk}} = (\alpha^{p^d})^{p^{d(k-1)}} = \alpha^{p^{d(k-1)}} = (\alpha^{p^d})^{p^{d(k-2)}} = \ldots = \alpha,$$

and therefore α is a root of $x^{p^n} - x$. Hence $x^{p^n} - x$ is divisible by all the factors $x - \alpha$ of f_d and so by f_d. Conversely, if f_d divides $x^{p^n} - x$, a root of f_d is also a root of of $x^{p^n} - x$, so that the splitting field F' of f_d is a subfield of F, the splitting field of $x^{p^n} - x$. Hence F is a vector space over F', and as such has $(p^d)^k$ elements, for some k. It follows $p^n = |F| = (p^d)^k$, from which $n = dk$, and d is a divisor of n. Finally, $x^{p^n} - x$ has no multiple factors. $\diamond$

Example. Consider $x^8 - x$. Here $p^n = 2^3$, so $d = 1$ or 3. The polynomials of degree 1 over Z_2 are x and $x+1$, the irreducible ones of degree 3 are $x^3 + x + 1$ and $x^3 + x^2 + 1$. It easy to see that the product of these four polynomials is $x^8 - x$.

Corollary 4.10. *If I_d is the number of monic irreducible polynomials of degree $d|n$ over Z_p then*

$$p^n = \sum_{d|n} d I_d. \tag{4.9}$$

Proof. Consider the degrees of the two sides of (4.8). $\diamond$

4.4 Cyclotomic polynomials

Let K be a field of characteristic p or zero, and consider the polynomial $x^n - 1$ over K. Its roots are the n-th roots of the unit of K; these are all distinct if the characteristic of K does not divide n. The set of these roots is a finite subgroup U of the splitting field of $x^n - 1$, which has order n if $p \nmid n$ (if $p|n$, let $n = p^k r$, $p \nmid r$; then $x^n - 1 = (x^r - 1)^{p^k}$, and the group has order r). By the argument of the previous section U is cyclic, generated by a primitive

n-th root w, and these primitive roots are $\varphi(n)$ in number (if $p|n$, there are no primitive n-th roots). So the polynomial $x^n - 1$ splits into linear factors:

$$x^n - 1 = \prod_{i=1}^{n} (x - w^i).$$

Collecting the primitive w^i we have a polynomial of degree $\varphi(n)$, which we denote $\varphi_n(x)$:

$$\varphi_n(x) = \prod_{\substack{k \\ \gcd(k,n)=1}} (x - w^k).$$

This polynomial is the *n-th cyclotomic polynomial*. If d is a divisor of n, $w^{\frac{n}{d}}$ is a primitive d-th root of unity. Define $\varphi_d(x)$ as the product of the linear factors corresponding to the d-th roots. It follows:

$$x^n - 1 = \prod_{d|n} \varphi_d(x). \tag{4.10}$$

Examples. 1. If $K = Q$, the rational field, then

$$x^4 - 1 = (x - 1)(x + 1)(x - i)(x + i).$$

The primitive fourth roots of unity are i and $-i$, so

$$\varphi_4(x) = (x - i)(x + i) = x^2 + 1.$$

The unique primitive second root of unity is -1, so

$$\varphi_2(x) = x + 1,$$

and the unique first root is 1, so

$$\varphi_1(x) = x - 1.$$

A primitive n-th root of unity over Q is the complex number $e^{\frac{2\pi i}{n}} = \cos(\frac{2\pi}{n}) + i\sin(\frac{2\pi}{n})$. The other roots are the numbers $e^{\frac{2\pi i}{n}k}$, $k = 2, \ldots, n$, and the primitive ones are those for which $\gcd(k, n) = 1$.

2. If $K = Z_p$ and $n = p - 1$ all the non zero elements are $(p - 1)$-st roots of unity (Corollary 4.5). For $p = 11$, the tenth primitive roots are 2, $8 = 2^3$, $7 = 2^7$ and $6 = 2^9$ mod 11, so

$$x^{10} - 1 = \prod_{k=1}^{10} (x - k),$$

$$\varphi_{10}(x) = (x - 2)(x - 8)(x - 7)(x - 6) = x^4 - x^3 + x^2 - x + 1,$$
$$\varphi_5(x) = (x - 4)(x - 5)(x - 9)(x - 3) = x^4 + x^3 + x^2 + x + 1,$$
$$\varphi_2(x) = x - 10 = x + 1,$$
$$\varphi_1(x) = x - 1.$$

3. Let $K = F_9$; its elements may be represented by the polynomials of degree 0 or 1 over Z_3. By choosing an irreducible polynomial $p(x)$ of degree 2 over Z_3, the product in F_9 is the usual product of polynomials followed by reduction modulo $p(x)$. For example, let $p(x) = x^2 + 1$; then all the elements of F_9 are obtained as powers of $x + 1$:

$$(x+1)^2 = x^2 + 1 + 2x \equiv 2x, (x+1)^3 \equiv 2x + 1, (x+1)^4 \equiv 2, \ldots, (x+1)^8 \equiv 1,$$

where the congruences are modulo $x^2 + 1$. Hence $x + 1$ is a primitive element whose minimal polynomial is $x^2 + x + 2$; the other ones are $2x + 1, 2x + 2$ and $x + 2$. (Note that the two roots x and $-x$ of $p(x)$ are not primitive elements of F_9.) From this follows that

$$\varphi_8(y) = (y-x-1)(y-2x-1)(y-2x-2)(y-x-2) = y^4 + (x^2+1)y^2 + x^4 + x^2 + 1,$$

which equals $y^4 + 1$ modulo $x^2 + 1$.

Remark. If K is a finite extension of a field F, the *primitive element theorem* states that there exists an element α in K such that $K = F(\alpha)$, that is, such that all the elements of K are rational functions of α with coefficients in F. The element α is said to be *primitive*. If K is a finite field we have called "primitive" an element $\alpha \in K$ that generates the multiplicative group of K; hence such an α is primitive also in the new sense because the powers of α are rational functions of α. The converse is false: in Example 3 above, the polynomial $f(x) = x$ is primitive in the new sense (every element of F_9 is a polynomial in x, hence a rational function of x) but not all elements of F_9 are powers of x (the powers of x are x, -1, $-x$, 1, and make up a cyclic group of order 4).

Let us now consider the rational field Q.

Theorem 4.13. *The cyclotomic polynomial $\varphi_n(x)$ over Q has integer coefficients.*

Proof. Induction on n. If $n = 1$, $\varphi_1(x) = x - 1$, which has integer coefficients. Assume the theorem true for $\varphi_m(x)$, $m < n$. From (4.10) we have

$$x^n - 1 = \prod_{\substack{d|n \\ d<n}} \varphi_d(x) \cdot \varphi_n(x).$$

By induction, the polynomials $\varphi_d(x)$ for $d < n$ have integer coefficients, and therefore so does their product. Moreover, the $\varphi_d(x)$ being monic, so is their product. It follows that the quotient of $x^n - 1$ by this product has integer coefficients, and this quotient is precisely $\varphi_n(x)$. $\diamond$

Remark. In a field of characteristic $p > 0$, this theorem holds if the elements of the base field Z_p, that is the multiples of 1, are considered as integers. In particular, factorisation (4.10) also holds for any finite field.

Formula (4.10) may be considered as a formula that allows the recursive calculation of the polynomials $\varphi_n(x)$. Indeed, if the $\varphi_d(x)$ with $d|n$ and $d < n$ are known, $\varphi_n(x)$ is obtained as a quotient of $x^n - 1$ by the product of the $\varphi_d(x)$. Hence we have $\varphi_1(x) = x - 1$, and for $n = p$, a prime,

$$\varphi_p(x) = \frac{x^p - 1}{x - 1} = x^{p-1} + x^{p-2} + \cdots + x + 1.$$

A few more expressions of $\varphi_n(x)$:

$$\varphi_4(x) = \frac{x^4 - 1}{(x-1)(x+1)} = x^2 + 1,$$

$$\varphi_6(x) = \frac{x^6 - 1}{(x-1)(x+1)(x^2 + x + 1)} = x^2 - x + 1,$$

$$\varphi_8(x) = x^4 + 1,$$

$$\varphi_9(x) = x^6 + x^3 + 1,$$

$$\varphi_{10}(x) = x^4 - x^3 + x^2 - x + 1,$$

$$\varphi_{12}(x) = x^4 - x^2 + 1.$$

Theorem 4.14. *The cyclotomic polynomial $\varphi_n(x)$ is irreducible over Q.*

Proof. Let us prove that the cyclotomic polynomial is the minimal polynomial of a primitive n-th root of unity w (and so is irreducible). Let $f(x)$ be the minimal polynomial of w; if we show that all the primitive n-th roots of unity are roots of $f(x)$ then we will have $\partial f(x) \geq \varphi_n(x)$, and therefore the equality $f(x) = \varphi_n(x)$ because $f(x)|\varphi_n(x)$.

The polynomial $f(x)$ divides $x^n - 1$; let $x^n - 1 = f(x)h(x)$, where both $f(x)$ and $h(x)$ are monic and integer (Gauss lemma). Let us first prove that if p is a prime that does not divide n, then w^p is a root of $f(x)$; if not, it must be a root of $h(x)$, $h(w^p) = 0$, and so w is a root of $h(x^p)$. Then $h(x^p) = f(x)g(x)$ for some $g(x)$, which is also integer because $f(x)$ is integer and monic. Taking everything modulo p we have (Theorem 4.8) $h(x)^p \equiv h(x^p) = f(x)g(x) \mod p$. If $q(x)$ is an irreducible factor of $f(x)$ in $Z_p(x)$, then $q(x)$ divides $h(x)$ in $Z_p(x)$ so that, again in $Z_p(x)$, $q(x)^2$ divides $x^n - 1$. But then $x^n - 1$ has a multiple factor, a contradiction because the derivative in $Z_p(x)$ is nx^{n-1}, which is non zero because $p \nmid n$, and therefore $x^n - 1$ is coprime with its derivative. Hence w^p is a root of $f(x)$.

Now, a primitive root has the form w^k, with $\gcd(k, n) = 1$, and therefore $k = p_1 p_2 \cdots p_t$, a product of primes not dividing n. We have just seen that $f(w^{p_1}) = 0$; by repeating the argument with w^{p_1} in place of w we see that $f(w^{p_1 p_2}) = 0$, and going on in this way we obtain $f(w^k) = 0$.   ◇

Remark. The fact that $f(w^k) = 0$ may be proved using Dirichlet's theorem stating that, if k and n are two coprime integers, then the arithmetic progression $k + hn$, $h = 1, 2, \ldots$, contains infinitely many primes. With w^k a primitive root, and hence with $\gcd(k,n) = 1$, we have, for infinitely many primes p, $w^p = w^{k+hn} = w^k \cdot w^{hn} = w^k$. But from $f(w) = 0$ it follows $0 = f(w)^p \equiv f(w^p) = f(w^k)$ mod p, hence the number $f(w^k)$ is zero modulo infinitely many primes and therefore is zero.

However, the cyclotomic polynomial may split over a finite field. We have seen that $\varphi_8(x) = x^4 + 1$ splits over Z_p for every p. In this connection we state without proof the following general theorem: if $p \nmid n$, and $q = p^n$, the polynomial $\varphi_n(x)$ splits over F_q into the product of $\varphi(n)/m$ irreducible factors of degree m, where m is the order of q mod n.

We close this section by remarking that taking the degrees of the two sides of (4.10) we obtain

$$n = \sum_{d|n} \varphi(d). \tag{4.11}$$

This may also be seen from the fact that the n-th roots of unity, that are n in number, divide up into primitive d-th roots, for every d that divides n. (See also the discussion before Theorem 4.11.)

As is the case of the cyclotomic polynomials, (4.11) may be considered as a formula that allows one to compute $\varphi(n)$ in a recursive way. Indeed, if one knows $\varphi(d)$ for $d|n$ and $d < n$ we can calculate $\varphi(n)$ as

$$\varphi(n) = n - \sum_{\substack{d|n \\ d<n}} \varphi(d).$$

Example. Let $n = 15$. Then by (4.11)

$$15 = \varphi(1) + \varphi(3) + \varphi(5) + \varphi(15) = 1 + 2 + 4 + \varphi(15),$$

and therefore $\varphi(15) = 15 - 7 = 8$. Indeed, the numbers less than 15 and coprime with 15 are the eight integers 1, 2, 4, 7, 8, 11, 13, and 14.

4.5 Modular greatest common divisors

When computing the greatest common divisor d of two integer polynomials f and g it may happen that d has coefficients that are greater in modulus than those of f or g, as the following example shows:[2]

$$f = x^3 + x^2 - x - 1 = (x+1)^2(x-1),$$
$$g = x^4 + x^3 + x + 1 = (x+1)^2(x^2 - x - 1),$$
$$d = x^2 + 2x + 1 = (x+1)^2.$$

Analogously, for two polynomials with gcd equal to 1 the result of the Euclidean algorithm may be an integer much greater than 1; this complicates

---

[2] Due to Davenport and Trager, and mentioned in [DST], page 131.

the procedure needlessly. This does not happen if the coefficients are taken modulo a prime p, because then the numbers that occur are bounded by $p-1$ (or by $(p-1)/2$, if the representation is balanced). However, it may happen that the gcd mod p is not equal to that over the integers. In this section we propose to briefly discuss this problem.

Let f and g be two integer polynomials, p a prime number, and let f_p and g_p be the polynomials obtained by taking the coefficients modulo p. Let us see how the two gcds

$$\gcd(f,g)_p \quad \text{and} \quad \gcd(f_p,g_p)$$

are related. In general, they are different. For instance, if $f = x + 1$ and $g = x - 1$ then $\gcd(f,g) = 1$ in $Z[x]$ and so also $\gcd(f,g)_2 = 1$. But in $Z_2[x]$ the two polynomials are equal and therefore $\gcd(f_2,g_2) = x + 1$. In general, for the degrees of the two gcds we have:

Theorem 4.15. $\partial\gcd(f_p,g_p) \geq \partial\gcd(f,g)_p$.

Proof. Let $f = \gcd(f,g)h$, $g = \gcd(f,g)k$, from which $f_p = \gcd(f,g)_p h_p$ and $g_p = \gcd(f,g)_p k_p$. It follows that $\gcd(f,g)_p$ divides both f_p and g_p, and so also $\gcd(f_p,g_p)$. $\diamond$

If p does not divide the leading coefficients of f and g, then $\partial\gcd(f,g)_p = \partial\gcd(f,g)$ and therefore, by the theorem we have just seen, $\partial\gcd(f_p,g_p) \geq \partial\gcd(f,g)$. Hence, if f_p and g_p are coprime, i.e. $\partial(\gcd(f_p,g_p)) = 0$, the same happens for f and g in $Z[x]$.

The inequality $\partial\gcd(f_p,g_p) > \partial\gcd(f,g)_p$ may only hold for finitely many primes p, as the following theorem shows.

Theorem 4.16. Let $d = \gcd(f,g)$. If p does not divide the leading coefficients of f and g and does not divide the resultant $R(f/d,g/d)$, then

$$\partial\gcd(f_p,g_p) = \partial d_p = \partial d.$$

Proof. Let $f = dh$, $g = dk$. Then $f_p = d_p h_p$ and $g_p = d_p k_p$; moreover,

$$\gcd(f_p,g_p) = \gcd(d_p h_p, d_p k_p) = d_p\gcd(h_p,k_p)$$

$$= d_p\gcd(\frac{f_p}{d_p}, \frac{g_p}{d_p}) = d_p\gcd((\frac{f}{d})_p, (\frac{g}{d})_p)$$

(because $f_p/d_p = h_p = (f/d)_p$). Hence $\gcd(f_p,g_p) \neq d_p$ if and only if the $\gcd((f/d)_p,(g/d)_p)$ is not a constant, and therefore if and only if $R((\frac{f}{d})_p,(\frac{g}{d})_p) = 0$. But this resultant equals $R(f/d,g/d)_p$ (the resultant is the determinant of a matrix, which is a sum of products of the entries of the matrix), so $R = 0$ if and only if p divides $R(f/d,g/d)$. $\diamond$

The polynomials f/d and g/d being coprime, $R(f/d,g/d)$ is non zero, and therefore has a finite number of divisors. Hence the inequality $\partial\gcd(f_p,g_p) > \partial\gcd(f,g)_p$ may only hold for a finite number of primes p, i.e. those that divide $R(f/d,g/d)$.

4.6 Square-free form of a polynomial

Let $u = u(x)$ be a polynomial of degree $n > 0$, and let u_i be the product of all the factors of multiplicity i of u, $i = 1, 2, \ldots, n$ (if for some i there are no factors of this type we set $u_i = 1$). Thus we have

$$u = u_1 u_2^2 u_3^3 \cdots u_r^r, \tag{4.12}$$

where r is the highest multiplicity for a factor of u (so $u_{r+1} = \ldots = u_n = 1$). (4.12) is the *square-free form* of u; note that the polynomial $u_1 u_2 u_3 \cdots u_r$ (strictly speaking, this is the square-free polynomial) has the same roots of u but they are simple. We now describe an algorithm that allows us to determine the number r and the polynomials u_i. Assume for now that the characteristic of the field is zero. For u in the form (4.12) we have

$$u' = u_1' u_2^2 \cdots u_r^r + 2u_2' u_1 u_2 u_3^3 \cdots u_r^r + \cdots + r u_r' u_1 u_2^2 \cdots u_{r-1}^{r-1} u_r^{r-1},$$

$$d = \gcd(u, u') = u_2 u_3^2 \cdots u_r^{r-1}, \quad \frac{u}{d} = u_1 u_2 u_3 \cdots u_r,$$

$$\frac{u'}{d} = u_1' u_2 u_3 \cdots u_r + 2u_2' u_1 u_3 \cdots u_r + \cdots + i u_i' u_1 u_2 \cdots \hat{u}_i \cdots u_r + \cdots$$
$$+ r u_r' u_1 u_2 u_3 \cdots u_{r-1}$$

$$\left(\frac{u}{d}\right)' = u_1' u_2 u_3 \cdots u_r + u_2' u_1 u_3 \cdots u_r + \cdots + u_i' u_1 u_2 \cdots \hat{u}_i \cdots u_r + \cdots$$
$$+ u_r' u_1 u_2 u_3 \cdots u_{r-1}$$

(the prime denotes differentiation, and $\hat{u}_i$ means that the term u_i is missing). Multiplying the last equality by i and subtracting from the last but one, we have

$$\frac{u'}{d} - i\left(\frac{u}{d}\right)' = (1 - i)u_1' u_2 u_3 \cdots u_i \cdots u_r$$
$$+ (2 - i)u_2' u_1 u_3 \cdots u_i \cdots u_r + \cdots \tag{4.13}$$
$$- u_{i-1}' u_1 u_2 \cdots u_i \cdots u_r + u_{i+1}' u_1 u_2 \cdots u_i \cdots u_r$$
$$+ \cdots + (r - i)u_r' u_1 u_2 \cdots u_i \cdots u_{r-1}.$$

Now, if an irreducible polynomial $p(x)$ divides u/d, then it divides one of the u_k, and only one because $\gcd(u_k, u_j) = 1$ if $k \neq j$. If $k \neq i$, then $p(x)$ divides all the summands of (4.13) except $(k - i)u_k' u_1 u_2 \cdots \hat{u}_k \cdots u_r$, and therefore it cannot divide the sum. The only irreducible polynomials dividing u/d and (4.13) are those dividing u_i and, since u_i divides u/d and (4.13), it is their gcd:

$$u_i = \gcd\left(\frac{u}{d}, \frac{u'}{d} - i\left(\frac{u}{d}\right)'\right).$$

This yields the following algorithm for determining the u_i:

input: u, $n = \partial u$;
output: $u_1, u_2, \ldots, u_n$.
$d := \gcd(u, u')$;
$v := \frac{u}{d}$;
$w := \left(\frac{u}{d}\right)'$;
$z := \frac{u'}{d}$;
for i from 1 to n do:
$u_i := (v, z - iw)$.

The number of the multiple factors of maximal order is given by the integer r for which $u_r \neq 1$ and $u_i = 1$ for $i = r+1, \ldots, n$.

Example. Let $u = x^4 + 2x^3 - 2x - 1$. Then

$$d = (u, u') = (x+1)^2, \quad \frac{u}{d} = x^2 - 1, \quad \frac{u'}{d} = 4x - 2, \quad \left(\frac{u}{d}\right)' = 2x.$$

It follows

$$u_1 = (x^2 - 1, 4x - 2 - 1 \cdot 2x) = (x^2 - 1, 2(x - 1)) = x - 1,$$
$$u_2 = (x^2 - 1, 4x - 2 - 2 \cdot 2x) = (x^2 - 1, -2) = 1,$$
$$u_3 = (x^2 - 1, 4x - 2 - 3 \cdot 2x) = (x^2 - 1, -2(x + 1)) = x + 1,$$
$$u_4 = (x^2 - 1, 4x - 2 - 4 \cdot 2x) = (x^2 - 1, -2(2x - 1)) = 1,$$

and therefore $u = (x - 1) \cdot 1 \cdot (x + 1)^3$.

Remark. Since u may have many factors with the same multiplicity, the u_i are in general reducible, hence (4.12) is not in general the splitting into irreducible factors. It is such a splitting if and only if the u_i are irreducible.

If the characteristic of the field is $p > 0$ the algorithm we have just seen may not work. For instance, with $u = x^p + 1$ we have $u' = 0$, $d = u$, $v = 1$, $w = z = 0$ and the algorithm yields $u_i = 1$ for all i, whereas since $x^p + 1 = (x + 1)^p$ we have $r = p$ and $x^p + 1 = 1 \cdot 1 \cdots 1 \cdot (x + 1)^p$. One may then proceed as follows. We know that if $u' = 0$ then $u = g^p$ for a certain g; then we can apply the algorithm to g and raise to the p-th power the u_i thus obtained. (If also $g' = 0$, then one proceeds for g as for u.) In the case of $u = x^p + 1$ we have $u = g^p$ with $g = x + 1$. The algorithm obviously yields the unique factor $x + 1$, which raised to the p-th power gives the above decomposition.

If we are not interested in collecting the factors with the same multiplicity, but only in a factorisation into square-free factors we may proceed as follows:

1. If $\gcd(u, u') = 1$, then u is square-free.

2. Let $d = \gcd(u, u')$ with $\partial d > 0$:

   2a. if $d = u$, then $u' = 0$ and we are in characteristic p.

   Hence $u = g^p$. We restart with g in place of u;

2b. if $d \neq u$, d is a non trivial factor of u, and u/d has no multiple factors. Hence $u = d \cdot \frac{u}{d}$, and if d has multiple factors we proceed as above with d in place of u. In this way we reduce ourselves to $u = v_1 v_2 \cdots v_s$, with the v_i square-free.

For the polynomial of the previous example this method yields the factorisation $u = v_1 \cdot v_2 \cdot v_3 = (x+1)(x+1)(x^2-1)$.

Finally, we observe that for a square-free integer polynomial we have from Theorem 4.10 that the resultant $R(f, f')$ is non zero, and indeed this condition is also sufficient. However, the same polynomial may not be square-free if taken modulo a certain prime p (for example, x^2+1 is square-free over the integers but it is not square-free modulo 2, because it equals $(x+1)^2$). For the polynomial to remain square-free, the resultant $R(f, f')$ over the integers must remain non zero when taken modulo p, that is, it must not be divisible by p. (In the case of the example the resultant is 4.)

The result of this section shows that the factorisation of a polynomial reduces to that of a square-free polynomial.

Exercises

1. Prove that the polynomial $(x^2+x+1)(x^3-a)$, where a is an integer that is not a cube, has a linear factor mod p for all p but none over the integers. [*Hint:* if $p \equiv 1 \bmod 3$, $p-1 = 3k$ and $a^{p-1} = a^{3k} = (a^k)^3 \equiv 1 \bmod p$, i.e. a third root of unity is a root of the first factor. If $p \not\equiv 1 \bmod 3$, then every element is a cube ($p = 3k+2$ and $(a^{-k})^3 \equiv a \bmod p$).]

2. Let K be a field of characteristic zero and let $f = \prod_{i=1}^{n}(x-\alpha_i)^{k_i}$ in an extension $K' \supseteq K$. Prove that $\gcd(f, f') = \prod_{i=1}^{n}(x-\alpha_i)^{k_i-1}$.

3. Prove that the number of irreducible polynomials of degree 2 over F_q is $q(q-1)/2$.

4. Let $p > 2$. Prove that -1 is a square mod p if and only if $p \equiv 1 \bmod 4$.

5. Use Corollary 4.5 to prove Wilson's theorem $(p-1)! \equiv -1 \bmod p$.

6. Prove that 2 is a primitive element of Z_{13}. (It is known that $p = 4q+1$ with q prime, then 2 is primitive in Z_p.)

7. Let n, m and t be integers. Prove that $\gcd(t^n-1, t^m-1) = t^{\gcd(n,m)} - 1$. [*Hint:* observe that $(t^n-1) - t^{n-m}(t^m-1) = t^{n-m} - 1$, and proceed by induction on the maximum between n and m, recalling that for each triple of integers h, k and s one has $\gcd(h, k) = \gcd(h, k-hs)$.]

8. Prove that $\gcd(x^{p^n} - x, x^{p^m} - x) = x^{p^{(n,m)}} - x$. [*Hint:* divide by x and apply the previous exercise.]

9. Prove that:

i) $2^n - 1$ and $2^m - 1$ are coprime if and only if n and m are coprime;

ii) $2^n \equiv 2^m \bmod 2^k - 1$ if and only if $n \equiv m \bmod k$.

10. Prove that, if $n = p_1^{h_1} p_2^{h_2} \cdots p_k^{h_k}$,

$$\varphi(n) = n \sum_{d|n} \frac{\mu(d)}{d} = n \cdot \prod_{\substack{p|n \\ p \ distinct}} \left(1 - \frac{1}{p}\right) = \prod_{i=1}^{k} p_i^{h_i - 1}(p_i - 1).$$

11. Prove that $\varphi_{12}(x)$ splits over Z_p for every p.

12. Prove that, for a polynomial $u = u(x)$ of degree n, the polynomials q_i given recursively by $q_1 = u$, $q_{k+1} = \gcd(q_k, q_k')$, $k = 1, 2, \ldots, n-1$, yield the square-free form of u by setting $u_i = q_i q_{i+2}/q_{i+1}^2$, $i = 1, 2, \ldots, r-1$, where r is the smallest integer such that $q_{r+1} = 1$.

13. Let $f = x^2 + ax + b$ be a polynomial of degree 2. Prove that for every $n \geq 2$ there exists a unique polynomial of the form $g = x^n + cx + d$ that admits f as a factor. Generalise to the case in which f is of degree m. [*Hint*: let g be a polynomial of the required form, and let $g = fq + r$, $\partial r < 2$. Then $g - r$ is the required polynomial.]

14. Prove that for a prime $p > 2$ we have $x^p - x \equiv 0 \mod 2p$.

4.7 The Möbius function

Formula (4.11) expresses the integer n in terms of the Euler's function $\varphi(d)$ of its divisors d, formula (4.8) the polynomial $x^{p^n} - x$ as the product of all the irreducible polynomials $f_d(x)$ having a degree d that divide n, and formula (4.9) p^n in terms of the number I_d of these polynomials. We now propose to invert these formulas, and others of the same type, that is, to determine $\varphi(d)$, $f_d(x)$ and I_d in function of the left-hand sides of the formulas.

All these situations are concerned with the divisibility relation between integers, which is a partial order relation, i.e. a relation which is reflexive, antisymmetric and transitive. Consider now the more general case involving any partially ordered (p.o.) set, or poset, $\mathcal{P}$; we denote by "$\leq$" the order relation ($\mathcal{P}$ might be the set of the integers or of the reals w.r.t. the usual ordering, the power set of a set w.r.t. inclusion, etc.). If $x, y \in \mathcal{P}$, the *segment* with endpoints x and y is the set of the elements $a \in \mathcal{P}$ such that $x \leq a \leq y$; $\mathcal{P}$ is *locally finite* if every segment is a finite set. For instance, the set $\mathcal{N}$ of the natural numbers ordered by division is locally finite because an integer has a finite number of divisors. Moreover $\mathcal{N}$ has a first element, i.e. 1; this implies that the segment $1 \leq a \leq y$ is the set $\{a \in \mathcal{N} | a \leq y\}$.

Let now F be a field and $\mathcal{P}$ a locally finite poset, and consider the set of functions f defined on $\mathcal{P} \times \mathcal{P}$ with values in F and such that $f(x, y) = 0$ if $x \not\leq y$. These functions make up a vector space over F (the sum of two functions and the product of a function by a scalar are defined in an obvious way). By defining the product $f * g$ of two functions as

$$(f * g)(x, y) = \sum_{x \leq u \leq y} f(x, u)g(u, y),$$

(the sum makes sense because $\mathcal{P}$ is locally finite) the vector space becomes an algebra, the *incidence algebra* $\mathcal{A}(\mathcal{P})$ of $\mathcal{P}$. Among these functions is Kronecker delta δ:

$$\delta(x, y) = \begin{cases} 1, \text{ if } x = y; \\ 0, \text{ otherwise.} \end{cases}$$

Let f be any function; then

$$(f * \delta)(x, y) = \sum_{x \leq u \leq y} f(x, u)\delta(u, y),$$

and, since $\delta(u, y) = 0$ if $u \neq y$, only the terms $f(x, u)$ with $u = y$ remain, that is, $f(x, y)$. Therefore $f * \delta = f$, i.e. δ is the unit element of this algebra. If a function f admits an inverse g, then it must be

$$\sum_{x \leq u \leq y} f(x, u)g(u, y) = \delta(x, y); \qquad (4.14)$$

in particular, $f(x, x)g(x, x) = 1$, and therefore $f(x, x) \neq 0$. This condition is also sufficient. Indeed, let us define g by induction on the cardinality of the segment $x \leq u \leq y$. If it has cardinality 1 (it is formed only by x) we set $g(x, x) = 1/f(x, x)$. Assume g is defined for segments of cardinality n, and let the segment $x \leq u \leq y$ be of cardinality $n + 1$; then set

$$g(x, y) = -\frac{1}{f(x, x)} \sum_{x < u \leq y} f(x, u)g(u, y),$$

(which makes sense because the segment $u \leq z \leq y$ has cardinality at most n and so g is defined on it). Thus we have proved that a necessary and sufficient condition for f to be invertible is that $f(x, x) \neq 0$ for every $x \in \mathcal{P}$.

Consider now the *zeta function* defined as follows:

$$\zeta(x, y) = \begin{cases} 1, \text{ if } x \leq y; \\ 0, \text{ otherwise.} \end{cases}$$

Since $\zeta(x, x) = 1 \neq 0$ this function has an inverse, which is called the *Möbius function*, and is denoted by μ. It is determined by induction, in the same way as before, starting with $\mu(x, x) = 1$. Moreover, since

$$\sum_{x \leq u \leq y} \zeta(x, u)\mu(u, y) = \delta(x, y),$$

we have, for $x = y$, $\zeta(x, x)\mu(x, x) = \delta(x, x) = 1 \cdot \mu(x, x) = 1$, and for $x \neq y$, $\zeta(x, u) = 1$ and therefore $\sum_{x \leq u \leq y} \mu(u, y) = 0$. The Möbius function is then characterised by the property $\mu(x, x) = 1$ and by any one of the following:

$$\begin{aligned} &\sum_{x \leq u \leq y} \mu(u, y) = 0, \\ &\sum_{x \leq u \leq y} \mu(x, u) = 0, \\ &\mu(x, y) = -\sum_{x < u \leq y} \mu(u, y), \\ &\mu(x, y) = -\sum_{x \leq u < y} \mu(x, u) \end{aligned} \qquad (4.15)$$

($x < y$ means $x \leq y$ and $x \neq y$). Let us show that in the case of the natural numbers μ is the function defined as follows:

$$\mu(x,y) = \begin{cases} 1, & \text{if } y/x = 1; \\ (-1)^k, & \text{if } y/x = p_1 p_2 \cdots p_k \text{ distinct primes;} \\ 0, & \text{otherwise.} \end{cases}$$

Then μ is a function of the single variable y/x; we will denote it $\bar{\mu}$. We have $\mu(x,x) = 1$ and therefore the first of the above conditions is fulfilled. If $x|y$, $x \neq y$, and $y/x = p_1^{h_1} p_2^{h_2} \cdots p_k^{h_k}$, then

$$\sum_{x|u|y} \mu(x,u) = \sum_{1|\frac{u}{x}|\frac{y}{x}} \bar{\mu}(\frac{u}{x}) = \sum_{1|h|\frac{y}{x}} \bar{\mu}(h);$$

the hs that divide y/x and for which $\bar{\mu}(h) \neq 0$ are the products of distinct primes among those that divide y/x. Hence they are the p_i, the products $p_i p_j$, $p_i p_j p_t$, etc. The number of these primes is k, and on these the value of $\bar{\mu}$ is -1; the number of products of two primes is $\binom{k}{2}$, and on these the value of $\bar{\mu}$ is 1, etc. It follows that the above sum is

$$\sum_{j=0}^{k} (-1)^j \binom{k}{j} = 0$$

(recall the binomial formula $(x+y)^k = \sum_{j=0}^{k} \binom{k}{j} x^j y^{k-j}$ and set $x = -1$ and $y = 1$). Hence the function $\mu(x,y)$ satisfies the second equality in (4.15), and therefore it is actually the Möbius function on the natural numbers.

The interest of the function μ lies mainly in the fact that it allows the "inversion" of functions f defined on $\mathcal{P}$ with values in F, and indeed this was our goal at the outset of the present section. Let f be defined starting from a function g as follows: $f(x) = \sum_{y \leq x} g(y)$. We want to prove that g is then determined by f. For instance, consider the case of the natural numbers with the divisibility relation. Then $g(1) = f(1)$, $g(2) = f(2) - f(1)$, $g(3) = f(3) - f(1)$, ..., $g(6) = f(6) - f(3) - f(2) + f(1)$, etc.

Theorem 4.17 (Möbius inversion formula). *If f and g are functions defined on $\mathcal{P}$ with values in the field F, and if*

$$f(x) = \sum_{y \leq x} g(y),$$

then

$$g(n) = \sum_{y \leq x} f(y)\mu(y,x).$$

Proof. Let us replace the function f in the second equality by its expression given by the first one. We have

$$\sum_{y \leq x} f(y)\mu(y,x) = \sum_{y \leq x}\sum_{u \leq y} g(u)\mu(y,x) = \sum_{y \leq x}\sum_u g(u)\zeta(u,y)\mu(y,x),$$

where we have used the equality $\sum_{u \leq y} g(u) = \sum_u g(u)\zeta(u,y)$. After inverting the summation order, and keeping in mind that μ is the inverse of ζ, we have

$$\sum_u g(u)\sum_{y \leq x}\zeta(u,y)\mu(y,x) = \sum_u g(u)\delta(u,x) = g(x),$$

as required.   ◇

In the case of the natural numbers, Möbius inversion takes the form (we simply write μ instead of $\bar{\mu}$):

$$f(n) = \sum_{d|n} g(d) \Rightarrow g(n) = \sum_{d|n} f(d)\mu\left(\frac{n}{d}\right);$$

$g(n)$ is then the difference between the sum of the terms for which the value of $\mu(\frac{n}{d})$ is 1 and those for which it is -1.

In multiplicative form:

$$f(n) = \prod_{d|n} g(d) \Longrightarrow g(x) = \prod_{d|n} f(d)^{\mu(\frac{n}{d})}.$$

Examples. 1. With $f(n) = n$ and $g(d) = \varphi(d)$ we have, from (4.11),

$$\varphi(n) = \sum_{d|n} \mu(d)\cdot\frac{n}{d} = n\sum_{d|n}\frac{\mu(d)}{d}.$$

2. By inverting (4.9) we may calculate the number of monic irreducible polynomials over Z_p. With $f(n) = p^n$ and $g(d) = dI_d$ we have:

$$I_n = \frac{1}{n}\sum_{d|n}\mu(d)p^{\frac{n}{d}}.$$

So, for example, for $p = 2$ and $n = 4$,

$$I_4 = \frac{1}{4}(\mu(1)2^4 + \mu(2)2^2 + \mu(4)2) = \frac{1}{4}(2^4 - 2^2) = 3,$$

(the three irreducible polynomials of degree 4 over Z_2 are x^4+x+1, x^4+x^3+1 and $x^4 + x^3 + x^2 + x + 1$).

3. Using the multiplicative Möbius inversion, with $f(n) = x^{p^n} - x$ and $g(d) = f_d(x)$, we have from (4.8):

$$f_d(x) = \prod_{d|n}(x^{p^d} - x)^{\mu(\frac{n}{d})}.$$

For example, for $p = 2$ and $n = 4$,

$$1 \cdot (x^4 - x)^{-1}(x^{16} - x) = \frac{(x^{16} - x)}{x^4 - x} = x^{12} + x^9 + x^3 + 1$$

which is the product of the three irreducible polynomials of degree 4 over Z_2 seen above.

For the cyclotomic polynomial we have, from (4.10),

$$\varphi_n(x) = \prod_{d|n}(x^d - 1)^{\mu(\frac{n}{d})}.$$

For example, let us compute $\varphi_{12}(x)$. We have

$$\varphi_{12}(x) = \frac{(x^2 - 1)(x^{12} - 1)}{(x^4 - 1)(x^6 - 1)} = x^4 - x^2 + 1,$$

because $\mu(12) = \mu(4) = 0$, $\mu(6) = 1$, $\mu(3) = \mu(2) = -1$. The direct calculation of this polynomial implies the division of a polynomial of degree 14 by one of degree 10. To avoid such a tedious division, we may proceed as follows. We know that the result will be a polynomial of degree $14 - 10 = 4$ (we also know it because $\varphi(12) = 4$), so that, taking modulo x^k the polynomials that appear in the quotient, with $k > 4$, the result will be the same. Let $k = 5$; then we have $x^{12} - 1 = x^5 \cdot x^7 - 1 \equiv -1 \bmod x^5$. Similarly, $x^6 - 1 \equiv -1 \bmod x^5$, so that

$$\varphi_{12}(x) = \frac{1 - x^2}{1 - x^4} = \frac{1}{1 + x^2} \bmod x^5.$$

We know (Chapter 2) that $\frac{1}{1+x^2} = 1 - x^2 + x^4 - x^6 + \cdots$, and this series modulo x^5 is precisely the polynomial $x^4 - x^2 + 1$.

4.8 Berlekamp's method

We now consider a general method for factoring a polynomial $u(x)$ over Z_p. We may limit ourselves to the case in which the polynomial is square-free, either by considering the square-free form seen in Section 4.6 and factoring the single u_i's, or factoring $u/(u, u')$. Moreover, one can always reduce to the case of a monic polynomial: it is sufficient to substitute x/a for x, where a is the leading coefficient of $u(x)$, and then multiply by a^{n-1}, where n is the degree of $u(x)$. The polynomial to be factored is then the monic polynomial $g(x) = a^{n-1}u(x/a)$.

Example. Let $u(x) = 3x^2 + 4x + 1$. Proceeding as described we have:

$$g(x) = 3\left(3\left(\frac{x}{3}\right)^2 + 4\left(\frac{x}{3}\right) + 1\right) = x^2 + 4x + 3 = (x + 1)(x + 3).$$

Conversely, if we substitute $3x$ for x and divide by 3 we obtain $\frac{1}{3}(3x+1)(3x+3) = u(x)$.

The method we are about to explain is due to Elwyn Berlekamp (1967). It allows us to find all the irreducible factors of $u(x)$. Let

$$u(x) = p_1(x)p_2(x)\cdots p_r(x)$$

be the factorisation of a square-free polynomial $u(x)$ we seek. $u(x)$ being square-free, the $p_i(x)$s are relatively prime. We try first to find the factors of $u(x)$ that are not necessarily irreducible; this can be done using polynomials that are found as follows.

For each polynomial $f(x)$, and if $s \neq t$ are two elements of Z_p, the polynomials $f(x) - s$ and $f(x) - t$ are relatively prime (if a polynomial divides both, then it divides their difference $t - s$, which is a constant); a fortiori,

$$\gcd(u(x), f(x) - s) \text{ and } \gcd(u(x), f(x) - t)$$

are relatively prime, and since both divide $u(x)$ so does their product $u(x)$. From this follows that

$$\prod_{s \in Z_p} \gcd(u(x), f(x) - s) \text{ divides } u(x). \tag{4.16}$$

Let us now see for which polynomials $f(x)$, if any, (4.16) holds. This happens if and only if for each irreducible factor $p_i(x)$ of $u(x)$ there exists s such that $p_i(x)$ divides $f(x) - s$. Now, given any r-tuple $s_1, s_2, \ldots, s_r$ of elements of Z_p, not necessarily distinct, since the $p_i(x)$s are pairwise coprime the Chinese remainder theorem implies the existence of a polynomial $v(x)$ such that

$$v(x) \equiv s_i \bmod p_i(x), \; i = 1, 2, \ldots, r, \tag{4.17}$$

$v(x)$ being unique modulo the product $\prod_{i=1}^{r} p_i(x)$, that is, modulo $u(x)$. Then, for this $v(x)$, for every $p_i(x)$ there exists s_i such that $p_i(x)$ divides $v(x) - s_i$, and therefore $\gcd(u(x), v(x) - s_i)$. Together with what we have seen above we may conclude with the following theorem.

Theorem 4.18. *Let $v(x)$ be a polynomial satisfying (4.17). Then*

$$u(x) = \prod_{s \in Z_p} \gcd(u(x), v(x) - s). \tag{4.18}$$

If $u(x)$ is reducible and $\partial v(x) \geq 1$, then (4.18) is a non-trivial factorisation of $u(x)$. Indeed, the degree of each gcd is smaller than that of $u(x)$ because $\partial v(x) < \partial u(x)$, and not all can be of degree zero (i.e. constant) because their product is $u(x)$, which is not constant.

We will say that a polynomial $v(x)$ of degree greater or equal to 1 for which (4.18) holds is a *reducing polynomial* for $u(x)$.

We now show that a solution of the system of congruences (4.17) depending on the unknown polynomials $p_i(x)$ is equivalent to the solution of a single congruence, which depends on the product of the $p_i(x)$s, that is, on the (known) polynomial $u(x)$. By Fermat's theorem, $s_i^p \equiv s_i \bmod p$; it follows that $v(x)^p \equiv s_i^p \equiv s_i \equiv v(x) \bmod p_i(x)$, from which $v(x)^p \equiv v(x) \bmod p_i(x)$, for all i, and therefore

$$v(x)^p \equiv v(x) \bmod u(x). \tag{4.19}$$

Thus, (4.19) is a necessary condition for (4.17) to hold. But it is also sufficient: indeed, if $v(x)$ satisfies (4.19) then $u(x)$ divides $v(x)^p - v(x)$, which therefore is also divided by all the $p_i(x)$s. Since

$$v(x)^p - v(x) = (v(x) - 0)(v(x) - 1) \cdots (v(x) - (p - 1))$$

each $p_i(x)$ must divide one of the right-hand factors, $v(x) - s_i$, say. In this way each factor $p_i(x)$ of $u(x)$ determines an element s_i of Z_p such that (4.17) holds. We have proved

Theorem 4.19. *Let r be the number of irreducible factors of $u(x)$. Then an r-tuple $s_1, s_2, \ldots, s_r$ of elements of Z_p determines a polynomial $v(x)$ satisfying (4.19). Conversely, such a polynomial determines an r-tuple satisfying (4.17). The number of these r-tuples being p^r, there are exactly p^r solutions of (4.17) and therefore of (4.19).*

Example. Let $p = 5$, $u(x) = x^4 + 1$; then $\gcd(u(x), u'(x)) = 1$, and $u(x)$ is square-free. As our polynomial $v(x)$ let us take $v(x) = x^2$. We have $v(x)^5 - v(x) = x^{10} - x^2 = (x^4 + 1)(x^6 - x^2) \equiv 0 \bmod x^4 + 1$, and therefore $v(x)^5 \equiv v(x) \bmod u(x)$, that is, $v(x)$ satisfies (4.19). The $\gcd(x^4 + 1, x^2 - s)$ for $s = 0, 1, 4$ are equal to 1; $\gcd(x^4 + 1, x^2 - 2) = x^2 - 2$ and $\gcd(x^4 + 1, x^2 - 3) = x^2 - 3$. Hence, $u(x) = x^4 + 1 = 1 \cdot 1 \cdot (x^2 - 2) \cdot (x^2 - 3) \cdot 1$.

Theorem 4.19 suggests that we look for polynomials satisfying (4.19). We will determine them as fixed points of a linear transformation of the vector space of the polynomials of degree smaller than the degree n of $u(x)$ over Z_p. This space has dimension n, a basis being given by the monomials $1, x, x^2, \ldots, x^{n-1}$, and therefore contains p^n polynomials.

The mapping:

$$\varphi : f(x) \to f(x)^p \bmod u(x)$$

is a linear transformation of the space. Indeed,

$$\begin{aligned}
\varphi(af(x) + bg(x)) &= (af(x) + bg(x))^p = a^p f(x)^p + b^p g(x)^p \\
&= af(x)^p + bg(x)^p \\
&= a\varphi(f(x)) + b\varphi(g(x)),
\end{aligned}$$

keeping in mind that $a^p \equiv a \bmod p$, and the same for b. The polynomials satisfying (4.19) are the fixed points of this linear transformation. The argument used to show that φ is linear also shows that the polynomials satisfying (4.19) form a subspace.

In the basis $1, x, x^2, \ldots, x^{n-1}$, φ is represented by a matrix whose entries on the k-th row are the coefficients of the polynomial taken modulo $u(x)$, which is the image under φ of the monomial x^k. Therefore these entries are the coefficients of the remainder of the division of x^{pk} by $u(x)$:

$$Q = \begin{pmatrix} q_{0,0} & q_{0,1} & \cdots & q_{0,n-1} \\ q_{1,0} & q_{1,1} & \cdots & q_{1,n-1} \\ \vdots & \vdots & \ddots & \vdots \\ q_{k,0} & q_{k,1} & \cdots & q_{k,n-1} \\ \vdots & \vdots & \ddots & \vdots \\ q_{n-1,0} & q_{n-1,1} & \cdots & q_{n-1,n-1} \end{pmatrix},$$

where:

$$\varphi(x^k) = x^{pk} \equiv q_{k,0} + q_{k,1}x + \cdots + q_{k,n-1}x^{n-1} \bmod u(x),$$

for $k = 0, 1, \ldots, n-1$. (Note that the first row of such a matrix, the row that corresponds to $x^0 \bmod u(x)$, is always $1, 0, \ldots, 0$.)

If $f(x) = \sum_{k=0}^{n} a_k x^k$ is a polynomial of the space under consideration, its image under φ is

$$\varphi(f(x)) = \varphi\left(\sum_{k=0}^{n-1} a_k x^k\right) = \sum_{k=0}^{n-1} a_k \varphi(x^k) = \sum_{k=0}^{n-1} a_k (x^{pk} \bmod u(x))$$

$$= \sum_{k=0}^{n-1} a_k \sum_{i=0}^{n-1} q_{k,i} x^i = \sum_{k,i=0}^{n-1} a_k q_{k,i} x^i.$$

This shows that the coefficients of $\varphi(f(x))$ are obtained as the product of the vector whose components are the coefficients of $f(x)$ by the matrix Q. The polynomials $v(x)$ that satisfy (4.19) are the ones fixed by φ; if $v_0, v_1, \ldots, v_{n-1}$ are the coefficients of such a polynomial, then:

$$(v_0, v_1, \ldots, v_{n-1})Q = (v_0, v_1, \ldots, v_{n-1}). \tag{4.20}$$

Conversely, if a vector satisfies this equality, its components are the coefficients of a polynomial fixed by φ. We sum up this discussion in the following theorem.

Theorem 4.20. *A polynomial over Z_p satisfies (4.19) if and only if its coefficients are the components of an eigenvector of the matrix Q relative to the eigenvalue 1, in the order of the increasing powers of x. Moreover, the number of these eigenvectors is p^r (because this is the number of the polynomials satisfying (4.19)).*

Thus, the number r of irreducible factors of $u(x)$ equals the geometric multiplicity of the eigenvalue 1 of Q. We are now in a position to determine the number r. Let $v = (v_0, v_1, \ldots, v_{n-1})$; if for this vector v (4.20) holds, then $v(Q - I) = 0$, where I is the $n \times n$ identity matrix. Thus r is the dimension of the kernel of $Q - I$. If ρ is the rank $Q - I$, then $n = r + \rho$.

Theorem 4.21. *The number r of irreducible factors of $u(x)$ equals the dimension of the kernel of $Q - I$, so it equals $r = n - \rho$, where ρ is the rank of $Q - I$.*

Remark. We can never have $\rho = n$ because the first row of Q is $1, 0, \ldots, 0$, and therefore the first row of $Q - I$ consists of only zeros, so that its rank ρ is at most $n - 1$.

The matrix Q always admits the eigenvector $(1, 0, \ldots, 0)$ and all its multiples $(a, 0, \ldots, 0)$, $a \in Z_p$. These vectors are the coefficients of the polynomials of degree zero; this corresponds to the fact that the elements of Z_p, considered as polynomials of degree zero, satisfy (4.19) by Fermat's theorem. If these are the only eigenvectors of Q, then $r = 1$, and $u(x)$ is irreducible, and conversely. We have the following corollary.

Corollary 4.11. *The polynomial $u(x)$ is irreducible if and only if the only polynomials satisfying (4.19) are constants.*

This corollary may also be proved using (4.17) and (4.18). Actually, if a non-constant solution $v(x)$ of (4.17) exists, then none of the gcds appearing in (4.18) can be equal to $u(x)$, since $\partial v(x) < \partial u(x)$. In this case at least two gcds are non trivial, and therefore $u(x)$ is reducible. If all the solutions are constant then, for each of these $v(x)$, system (4.17) reduces to one congruence, the one in which s_i equals the constant $v(x)$, and therefore $u(x)$ coincides with its irreducible factor $p_i(x)$.

Finally, we observe that if $v(x)$ is constant, $v(x) = t$, then $\gcd(u(x), t - s)$ equals 1 if $s \neq t$, and equals $u(x)$ if $s = t$. The factorisation is then the trivial one $u(x) = 1 \cdot 1 \cdots 1 \cdot u(x) \cdot 1 \cdots 1$.

Corollary 4.12. *The polynomial $u(x)$ splits into linear factors over Z_p if and only if Q is the identity matrix, and therefore if and only if all the polynomials of degree less than n are $v(x)$s that reduce $u(x)$.*

Proof. The polynomial $u(x)$ splits into linear factors if and only if $r = n$, and this happens if and only if $\rho = 0$, that is $Q = I$. Moreover, $r = n$ means that $Ker(Q - I)$ contains p^n vectors, that is, all the polynomials of degree less than n. $\diamond$

The following corollary shows that if $v(x) = x$ is a polynomial that reduces $u(x)$, then all polynomials reduce $u(x)$.

Corollary 4.13. *The polynomial $u(x)$ splits into linear factors over Z_p if and only if $v(x) = x$ is a polynomial reducing $u(x)$.*

Proof. By (4.17), we have $x \equiv s_i \bmod p_i(x)$, so $p_i(x) = x - s_i$, and the s_is are distinct because the $p_i(x)$s are. Conversely, if $u(x)$ splits into linear factors, then $u(x) = \prod_{i=1}^{r}(x - s_i)$ and therefore $u(x)$ divides $x^p - x = \prod_{s \in Z_p}(x - s)$, so $x^p - x \equiv 0 \bmod u(x)$. By (4.19), $v(x) = x$ is a polynomial reducing $u(x)$. $\diamond$

Examples. 1. Let $u(x) = x^4 - x^2 + 1$ and $p = 5$. It is immediate that $\gcd(u(x), u'(x)) = 1$. Since the calculations are made modulo $u(x)$, set $u(x) = 0$, i.e. $x^4 = x^2 - 1$. Now we have:

$$x^0 = 1 = 1 + 0 \cdot x + 0 \cdot x^2 + 0 \cdot x^3;$$
$$x^5 = x^4 \cdot x = (x^2 - 1)x = x^3 - x = 0 - 1 \cdot x + 0 \cdot x^2 + 1 \cdot x^3;$$
$$x^{10} = x^5 \cdot x^5 = (x^3 - x)(x^3 - x) = x^6 - 2x^4 + x^2.$$

But $x^6 = x^4 \cdot x^2 = (x^2 - 1)x^2 = x^4 - x^2 = x^2 - 1 - x^2 = -1$, so that
$$x^{10} = -1 - 2x^2 + 2 + x^2 = 1 - x^2 = 1 + 0 \cdot x - 1 \cdot x^2 + 0 \cdot x^3;$$
$$x^{15} = x^{10} \cdot x^5 = (1 - x^2)x^5 = x^5 - x^7.$$

But $x^7 = x^5 \cdot x^2 = (x^3 - x)x^2 = x^5 - x^3 = (x^3 - x) - x^3 = -x$, and therefore
$$x^{15} = x^3 - x + x = x^3 = 0 + 0 \cdot x + 0 \cdot x^2 + 1 \cdot x^3.$$

Hence we have the matrix:

$$Q = \begin{pmatrix} 1 & 0 & 0 & 0 \\ 0 & -1 & 0 & 1 \\ 1 & 0 & -1 & 0 \\ 0 & 0 & 0 & 1 \end{pmatrix};$$

so

$$Q - I = \begin{pmatrix} 0 & 0 & 0 & 0 \\ 0 & -2 & 0 & 1 \\ 1 & 0 & -2 & 0 \\ 0 & 0 & 0 & 0 \end{pmatrix}.$$

Let us seek the vectors annihilated by $Q - I$:

$$(v_0, v_1, v_2, v_3)(Q - I) = (0, 0, 0, 0);$$

we have $(v_2, -2v_1, -2v_2, v_1) = (0, 0, 0, 0)$, from which $v_1 = v_2 = 0$; the other two components are arbitrary. Therefore, the vectors annihilated by $Q - I$ are those of type $(a, 0, 0, b)$, with a and b arbitrary in Z_5, and only these. Hence they are $5^2 = 25$ in number, and a basis of the subspace they form (i.e., the kernel of $Q - I$), is given by $(1, 0, 0, 0)$ and $(0, 0, 0, 1)$. Thus $\dim(Ker(Q - I)) = 2$, $r = n - 2 = 4 - 2 = 2$, so that $x^4 - x^2 + 1$ splits into two irreducible factors over Z_5. In order to find these two factors let us take a polynomial of $Ker(Q - I)$, $v(x) = x^3$ say. By calculating $\gcd(x^4 - x^2 + 1, x^3 - s)$, with $s = 0, 1, 2, 3, 4$, we find that two of them, those for which $s = 2$ and $s = 3$, are different from 1, while the other ones are equal to 1. Hence,

$$u(x) = 1 \cdot 1 \cdot (2x^2 + x - 2) \cdot (3x^2 + x + 2) \cdot 1.$$

By dividing the first factor by 2 and the second by 3 (this amounts to dividing $u(x)$ by 6, that is by 1), we have two monic factors, $x^2 + 3x - 1$ and $x^2 + 2x - 1$.

2. Consider the polynomial $u(x) = x^{p-1} + 1$ over Z_p, $p > 2$. Since $u'(x) = (p-1)x^{p-2}$ we have $\gcd(u(x), u'(x)) = 1$. Moreover,

$$x^{pk} = (x^{p-1})^k \cdot x^k \equiv (-1)^k \cdot x^k,$$

so that the matrix is the diagonal matrix:

$$Q = \mathrm{diag}(1, -1, 1, \ldots, 1, -1).$$

It follows $Q - I = \mathrm{diag}(0, -2, 0, \ldots, 0, -2)$, and the rank of $Q - I$, which in this case is the number of non-zero entries on the diagonal, is $\rho = \frac{p-1}{2}$. Since $n = p - 1$, the number of irreducible factors of the polynomial is $r = n - \rho = \frac{p-1}{2}$.

For a vector of the kernel of $Q - I$, we have $(0, -2v_1, 0, -2v_3, \ldots, 0, -2v_{p-2}) = (0, 0, \ldots, 0)$, and therefore $v_1 = v_3 = \ldots = v_{p-2} = 0$, and the other components are arbitrary. A basis of the kernel of $Q - I$ is given by the vectors having 1 at entries in positions $0, 2, \ldots, p-2$ and zero elsewhere, which therefore correspond to the polynomials $v(x) = 1, x^2, x^4, \ldots, x^{p-3}$.

Thus, for instance, with $p = 5$ and $v(x) = x^2$, we have $\gcd(x^4 + 1, x^2 - s) = 1$ for $s = 0, 1, 4$, and equal to $x^2 - 2$ and $x^2 - 3$ for $s = 2$ and 3, respectively, and these two factors are irreducible. Note that the polynomial $v(x) = x^2$ corresponds to the pair of elements of Z_5 given by 2 and 3; in fact, $x^2 \equiv 2 \bmod (x^2 - 2)$ and $x^2 \equiv 3 \bmod (x^2 - 3)$.

3. If $u(x)$ is not square-free, it may happen that $r \neq n - \rho$. For example, with $u(x) = x^2 + 1$ and $p = 2$ we have $Q = \begin{pmatrix} 1 & 0 \\ 1 & 0 \end{pmatrix}$ and $Q - I = \begin{pmatrix} 0 & 0 \\ 1 & 1 \end{pmatrix}$, and $\rho(Q - I) = 1$. $n - \rho = 2 - 1 = 1$ and $u(x)$ would be irreducible, whereas $u(x) = (x + 1)^2$.

For some $v(x)$, factorisation (4.18) may not be complete. Indeed, if for such a $v(x)$ two of the s_i of (4.17) are equal, $s_i = s_j = s$, then $\gcd(u(x), v(x) - s)$ will be a polynomial $w(x)$ divisible by $p_i(x)$ and $p_j(x)$ (and possibly other $p_k(x)$s), but in which these two factors are not evident because $w(x)$ appears as the product of its own factors. We shall then say that $v(x)$ *does not separate* the factors $p_i(x)$ and $p_j(x)$ of $u(x)$. In this case (4.18) will have the form $u(x) = w_1(x) \cdot w_2(x) \cdots w_t(x)$, with $t < r$ and one of the $w_i(x)$s equal to $w(x)$.

Corollary 4.14. *The factorisation of $u(x)$ given by (4.18) is complete if and only if $v(x)$ satisfies (4.17) with the s_i all distinct.*

Example. Let us see now an example of a polynomial $v(x)$ that does not separate all the factors of $u(x)$. Let $u(x) = x^3 - x^2 + x - 1$ with $p = 5$. This polynomial splits into linear factors over Z_5 as $(x - 1)(x - 2)(x - 3)$, as is easily seen, so that the matrix Q is the identity matrix. The polynomials of degree 2 belong to the kernel, i.e. they are $v(x)$s for which (4.20) holds. Let us consider $v(x) = 3x^2 + x + 2$; then the gcds equal 1 for $s = 0, 3, 4$, and equal $x^2 - 2x + 2$ and $x - 3$, for $s = 1$ and 2, respectively. The two factors $x - 1$ and $x - 2$ are merged in the product $x^2 - 2x + 2$, and this happens because $v(x)$ is congruent modulo $x - 1$ and modulo $x - 2$ to the same element of Z_5, that is to 1.

If the factorisation is not complete, take one of the obtained factors $w(x)$ and another $v(x)$, and proceed as before. In the next theorem we will see that all the factors $p_i(x)$ can be obtained in this way. Moreover, in choosing the $v(x)$s we may restrict ourselves to a particular set of polynomials.

Let $v^{(k)} = (a_{k_0}, a_{k_1}, \ldots, a_{k_{n-1}})$, $k = 1, 2, \ldots, r$, be a basis for the kernel of $Q - I$, and let $v_k(x) = a_{k_0} + a_{k_1} x + \cdots + a_{k_{n-1}} x^{n-1}$, that is, the polynomial whose coefficients are the components of $v^{(k)}$.

Theorem 4.22. *Given two irreducible factors $p_i(x)$ and $p_j(x)$ of $u(x)$, there exists one of the above defined polynomials $v_k(x)$ that separates them.*

Proof. If none of the $v_k(x)$ separates the two factors, then each of them is congruent to the same element of Z_p modulo both factors. That is, for each k there exists s_k such that $v_k(x) \equiv s_k \bmod p_i(x)$ and $v_k(x) \equiv s_k \bmod p_j(x)$. Now, a vector v of the kernel of $Q - I$ is a linear combination of the $v^{(k)}$, so that the $v(x)$ satisfying (4.17) is a linear combination of the polynomials $v_k(x)$: $v(x) = \sum_{k=1}^{r} c_k v_k(x)$. It follows $v(x) = \sum_{k=1}^{r} c_k v_k(x) \equiv \sum_{k=1}^{r} c_k s_k = s \bmod p_i(x)$, and similarly $v(x) \equiv s \bmod p_j(x)$.

Hence a $v(x)$ satisfying (4.17) is congruent modulo $p_i(x)$ and $p_j(x)$ to one and the same $s \in Z_p$. However, we have seen (Theorem 4.19) that an r-tuple $\{s_i\}$ gives rise to a polynomial $v(x)$ that satisfies (4.17), and an r-tuple for which $s_i \neq s_j$ holds is such that the corresponding $v(x)$ is congruent to two distinct elements of Z_p (which are precisely the s_i and s_j above). $\Diamond$

Let us see now what is the probability that a polynomial $v(x)$ satisfying (4.19) and chosen at random separates all the factors. By Corollary 4.14, it is the probability that r elements of Z_p chosen at random be all distinct. The r-tuples of distinct elements of Z_p, $r < p$, are

$$p(p-1) \cdots (p - (p - (r+1))) = \frac{p!}{(p-r)!},$$

in number (and this is the number of favourable outcomes), and since we have a total number of p^r r-tuples (possible choices), the probability we are looking for is

$$P = \frac{p!}{(p-r)! p^r}.$$

Now, if r is much smaller than p, this probability is approximately equal to 1. For instance, with $r = 2$, we have: for $p = 7$, $P = 6/7 \sim 0.86$; for $p = 11$, $P = 10/11 \sim 0.9$; and for $p = 17$, $P = 16/17 \sim 0.94$.

Summing up, Berlekamp's method consists of the following steps:

1. Reduce to the case in which $u(x)$ is square-free (as seen in § 4.6).

2. Form matrices Q and $Q - I$.

3. The rank ρ of $Q - I$ allows one to determine the number of irreducible factors, which is given by $r = n - \rho$. If $r = 1$, then $u(x)$ is irreducible, and the procedure ends, otherwise go to step 4.

4. The kernel of $Q - I$ ha dimension $r > 1$. Determine a basis v^k, $k = 1, 2, \ldots, r$, and the corresponding polynomials $v_k(x)$. Let $v_1(x) = 1$.

5. Compute $\gcd(u(x), v_k(x) - s)$ with one of the $v_k(x)$s, $k \geq 2$, as s varies in Z_p. For some s, this gcd is a non trivial factor of $u(x)$ (but not necessarily one of the $p_i(x)$s) since its degree is smaller than that of $u(x)$, and it cannot be 1 for all ss because by (4.18) the product of all the gcds is $u(x)$.

6. Let $u(x) = w_1(x)w_2(x)\cdots w_t(x)$ be the factorisation obtained at step 5. If $t = r$, the $w_i(x)$s coincide with the $p_i(x)$s and the procedure ends. Otherwise, repeat step 5 with a different $v_k(x)$ and with the various factors $w_i(x)$ in place of $u(x)$. By Theorem 4.22 this procedure yields all the factors $p_i(x)$.

4.8.1 Reducing the calculation of the gcds: the Zassenhaus-Cantor method

When p is very large with respect to n, the number of gcds one should compute is very large. Let us see how to find some factors of $u(x)$ (not necessarily one of the $p_i(x)$s) by replacing step 5 as follows. Note that

$$v(x)^p - v(x) = v(x)\left(v(x)^{\frac{p-1}{2}} + 1\right)\left(v(x)^{\frac{p-1}{2}} - 1\right).$$

Since $u(x)$ divides the left-hand side, it is likely that some factor of $u(x)$ is also a factor of $v(x)^{\frac{p-1}{2}} - 1$. Instead of $\gcd(u(x), v(x) - s)$, let us calculate $\gcd(u(x), v(x)^{\frac{p-1}{2}} - 1)$: this will be a non trivial factor of $u(x)$ with a certain probability, which we now compute. First, let us consider the probability that it is a trivial factor, $\gcd = u(x)$ or 1.

1. $\gcd = u(x)$. This happens if and only if all the $p_i(x)$s divide $v(x)^{\frac{p-1}{2}} - 1$. Now from (4.17) we have $v(x)^{\frac{p-1}{2}} \equiv s_i^{\frac{p-1}{2}}$, and therefore $p_i(x)$ divides $v(x)^{\frac{p-1}{2}} - 1$ if and only if $s_i^{\frac{p-1}{2}} \equiv 1 \bmod p$, i.e. when s_i is a solution of $x^{\frac{p-1}{2}} - 1 = 0$ in Z_p. This equation has exactly $\frac{p-1}{2}$ solutions in Z_p. Indeed, by dividing $x^p - x = \prod_{i=0}^{p-1}(x - i)$ by x we have $x^{p-1} - 1 = \prod_{i=1}^{p-1}(x - i)$. On the other hand, $x^{p-1} - 1 = (x^{\frac{p-1}{2}} - 1)(x^{\frac{p-1}{2}} + 1)$, so that $x - i$ divides one of the two factors, for all $i \neq 0$. However, exactly half of the $x - i$ divide $x^{\frac{p-1}{2}} - 1$: they cannot be more because the polynomial $x^{\frac{p-1}{2}} - 1$,

which is of degree $\frac{p-1}{2}$, cannot have more than $\frac{p-1}{2}$ roots, and they cannot be less otherwise the other factor would have more than $\frac{p-1}{2}$ roots. Hence, the favourable cases for a given $p_i(x)$ to divide $v(x)^{\frac{p-1}{2}} - 1$ are $\frac{p-1}{2}$, and that all the $p_i(x)$s divide it are $\left(\frac{p-1}{2}\right)^r$. There being p^r choices for $v(x)$, the probability that $\gcd = u(x)$ is

$$\frac{\left(\frac{p-1}{2}\right)^r}{p^r} = \left(\frac{p-1}{2p}\right)^r.$$

2. $\gcd = 1$. By the above the favourable cases, in which $p_i(x)$ does not divide $v(x)^{\frac{p-1}{2}} - 1$, are $p - \frac{p-1}{2} = \frac{p+1}{2}$, and therefore the probability that none divides is

$$\frac{\left(\frac{p+1}{2}\right)^r}{p^r} = \left(\frac{p+1}{2p}\right)^r.$$

Hence the probability that $\gcd(u(x), v(x)^{\frac{p-1}{2}} - 1)$ is a non trivial factor of $u(x)$ is

$$1 - \left(\frac{p-1}{2p}\right)^r - \left(\frac{p+1}{2p}\right)^r.$$

For $r \geq 2$ and $p \geq 3$ this is greater than or equal to $4/9$.

One may then proceed as follows. Let us choose at random a polynomial

$$v(x) = a_1 v_1(x) + a_1 v_1(x) + \cdots + a_r v_r(x),$$

where the $v_k(x)$s are the polynomials of the basis and the a_is are chosen at random in Z_p, and let us calculate

$$g(x) = \gcd(u(x), v(x)^{\frac{p-1}{2}} - 1).$$

If $g(x) = 1$ take another $v(x)$, otherwise $u(x) = u_1(x)u_2(x)$, where $u_1(x) = g(x)$ and $u_2(x) = u(x)/g(x)$, and repeat the calculation of the gcd with $u_1(x)$ and $u_2(x)$ in place of $u(x)$.

4.8.2 Reducing the calculations of the gcds: the method of the resultant

Given $v(x)$, many of the $\gcd(u(x), v(x) - s)$s of step 5 equal 1, and therefore their calculation is useless (this certainly happens if p is much greater than the degree of $u(x)$ since the number of factors of $u(x)$ is at most n). How may one discover for which elements s of Z_p the gcd is 1? Here the theory of the resultant helps us. Indeed, $\gcd \neq 1$ if and only if the resultant $R(u(x), v(x) - s)$ equals zero, i.e. if and only if s is a root of the polynomial $r(y) = R_x(u(x), v(x) - y)$ (which is of degree n). If $s \in Z_p$ is such a root, then $y - s$ divides $r(y)$; but $y - s$ also divides $y^p - y$, which is the product of all the $y - i$ as i varies in Z_p. It follows that $y - s$ divides $\gcd(r(y), y^p - y)$, which is a certain polynomial $g(y)$. Conversely, if $y - s$ divides $g(y)$, then s is a root of $r(y)$. Hence, the roots of $r(y)$ in Z_p are those of $g(y)$ (which has all its roots in Z_p).

What is the degree of $g(y)$? This polynomial only has simple roots, because this is the case for $y^p - y$, and therefore its degree equals the number of its roots. Moreover, if $s_1, s_2, \ldots, s_t$ are these roots, then they are also the roots of $r(y)$ in Z_p, and as such they give rise to factors $u_i(x) = \gcd(u(x), v(x) - s_i)$ of $u(x)$ which are non trivial (and distinct). And since, conversely, a non trivial gcd corresponds to one of the s_i, we conclude that the degree of $g(y)$ equals the number of factors in which $u(x)$ splits by means of $v(x)$. In particular, this degree is less than or equal to r.

Hence, the degree t of $g(y)$ tells us whether the factorisation obtained by means of $v(x)$ is the complete one, and this is the case if and only if $t = r$.

Remark. The three polynomials u, r and g may be equal. For this to happen, it is necessary and sufficient that u splits into linear factors over Z_p. Indeed, by what we have seen above, $\partial r = \partial u$ means that $v(x)$ splits u into $t = n$ factors. Conversely, if u splits into linear factors, by Corollary 4.13 $v(x) = x$ is a polynomial that reduces u, and with this polynomial $r(y) = R_x(u(x), x - y) = u(y)$. Moreover, since all the roots of u, and therefore of r, belong to Z_p, $g(y) = \gcd(r(y), y^p - y) = r(y)$.

Example. Let $p = 13$ and $u = x^8 + x^6 + 10x^4 + 10x^3 + 8x^2 + 2x + 8$. With the polynomial $v = x^6 + 5x^5 + 9x^4 + 5x^2 + 5x$ we have

$$r(y) = R(u, v - y) = y^8 - 6y^7 - y^6 + 5y^5 = (y - 2)^3 y^5.$$

Since we are interested in the roots of $r(y)$, let us consider the polynomial obtained by eliminating multiple factors, i.e. by dividing by $\gcd(r, r') = y^6 - 4y^5 + 4y^4$; we obtain $r_1 = y^2 - 2y$. Moreover,

$$\prod_{s \in Z_{13}} \gcd(u, v - s) = (x^5 + 5x^4 - 4x^3 + 5x + 5)(x^3 - 5x^2 + 4x - 1)$$

where the two factors correspond to $s = 0$ and $s = 2$. Then the polynomial $g(y)$ will have degree 2; indeed, $g(y) = \gcd(r_1, y^{13} - y) = y^2 - 2y$, and the roots are 0 and 2, which correspond to the two factors already found.

In order to calculate the coefficients of the powers of x taken modulo $u(x)$ that are needed to form the matrix Q we may use a relation that we now derive. Let

$$u(x) = x^n + u_{n-1}x^{n-1} + \cdots + u_1 x + u_0,$$

and let $x^k = a_{k,0} + a_{k,1}x + \cdots + a_{k,n-1}x^{n-1}$. Multiplying by x, we have $x^{k+1} = a_{k,0}x + a_{k,1}x^2 + \cdots + a_{k,n-1}x^n$. Now, $x^n \equiv -u_{n-1}x^{n-1} - \cdots - u_1 x - u_0 \mod u(x)$, which substituted in the previous one yields $x^{k+1} = a_{k+1,0} + a_{k+1,1}x + \cdots + a_{k+1,n-1}x^{n-1}$ where, and this is the promised relation,

$$a_{k+1,j} = a_{k,j-1} - a_{k,n-1}u_j \quad (a_{k,-1} = 0).$$

Remark. An important application of the factorisation modulo a prime p can be found in Galois theory. We have the following theorem, due to Dedekind: if p does not divide the discriminant of a polynomial $u(x)$, and in Z_p the polynomial splits into factors of degrees $n_1, n_2, \ldots, n_k$, then in the Galois group of $u(x)$ there exists a permutation whose cycles have lengths $n_1, n_2, \ldots, n_k$. For instance, consider the polynomial $u(x) = x^7 + 2x + 2$; its discriminant is -1, so all primes are acceptable. Let $p = 3$; then Berlekamp's method says that $u(x)$ has two factors over Z_3, and having no roots in Z_3, the two factors either are of degree $n_1 = 2$ and $n_2 = 5$, or $n_1 = 3$ and $n_2 = 4$ (it can be seen that the first case holds). Then by Dedekind's theorem the Galois group of $u(x)$ either contains a permutation that has a cycle of length 2 and one of length 5, or one having a cycle of length 3 and one of length 4. However, the only transitive ($u(x)$ is irreducible) subgroup of S_7 containing a permutation of one of the two types is the whole group S_7, hence this is the Galois group of $u(x)$. In particular, since the group S_7 is not soluble, the equation $u(x) = 0$ is not soluble by radicals.

4.8.3 The characteristic polynomial of matrix Q

As we have seen in Chapter 1, a polynomial $f(x)$ of degree less that of $u(x)$ can be written in a unique way as

$$f(x) \equiv f_1(x)L_1(x) + f_2(x)L_2(x) + \cdots + f_r(x)L_r(x) \bmod u(x),$$

where $L_1(x)$, $L_2(x)$, $\ldots$, $L_r(x)$ are the Lagrange polynomials relative to the polynomials $p_i(x)$, and where $f_i(x)$ is the remainder of the division of $f(x)$ by $p_i(x)$. The algebra A of the polynomials of degree less than n decomposes into the direct sum of subalgebras:

$$A = A_1 \oplus A_2 \oplus \cdots \oplus A_r,$$

with $A_k = \{f_k(x)L_k\} \simeq Z_p[x]/(p_k(x))$, $k = 1, 2, \ldots, r$. The polynomial $p_k(x)$ being irreducible, A_k is a field containing p^{n_k} elements, where $n_k = \partial p_k(x)$. The subalgebras A_k are invariant under the automorphism $\varphi : f(x) \to f(x)^p$ of $Z_p[x]/(u(x))$. Indeed,

$$f(x)^p \equiv f_1(x)^p L_1(x)^p + f_2(x)^p L_2(x)^p + \cdots + f_r(x)^p L_r(x)^p \bmod u(x),$$

and taking as $r_k(x)$ the remainder of the division of $f_k(x)^p$ by $p_k(x)$, the element $r_k(x)L_k(x)^p$ still belongs to A_k because, $L_k(x)$ being an idempotent, $L_k(x)^p \equiv L_k(x) \bmod u(x)$.

The study of the linear transformation φ can then be done on each of the components A_k. In other words, we may assume $u(x)$ irreducible and $A_k = A$ a field of order $q = p^n$. If $\alpha \in A$, we have $\varphi^n(\alpha) = \alpha^q = \alpha$, and therefore $\varphi^n(\alpha) = id$. Then the matrix Q of φ in the basis $1, x, \ldots, x^{n-1}$ is such that $Q^n = I$, so Q satisfies the polynomial $\lambda^n - 1$. We claim that this is the minimal polynomial of Q. Indeed, if Q satisfies a polynomial $\sum_{i=1}^{n-1} a_i \lambda^i$ of smaller degree, then $\sum_{i=1}^{n-1} a_i Q^i = 0$, and by applying this equality to the basis of the

x^i we have the homogeneous system

$$\begin{cases} a_0 + a_1 + \cdots + a_{n-1} = 0, \\ a_0 x + a_1 x^p + \cdots + a_{n-1} x^{p^{n-1}} = 0, \\ a_0 x^2 + a_1 x^{2p} + \cdots + a_{n-1} x^{2p^{n-1}} = 0, \\ \cdots \\ a_0 x^{n-1} + a_1 x^{(n-1)p} + \cdots + a_{n-1} x^{(n-1)p^{n-1}} = 0. \end{cases}$$

Setting $x_i = x^{p^i}$, the determinant of this system is the Vandermonde of the x_is, and since these are all distinct, the determinant is not zero, and the system only has trivial solutions. This shows that $\lambda^n - 1$ is the minimal polynomial of Q, and therefore, up to sign, also its characteristic polynomial:

$$\det(Q - \lambda I) = (-1)^n (\lambda^n - 1).$$

We may now go back to the general case. The A_k being invariant under φ, the matrix Q on A is block diagonal:

$$Q = \mathrm{diag}(C_1, C_2, \ldots, C_r),$$

where each block C_k is of dimension $n_k = \partial p_k(x)$. Then the characteristic polynomial of Q is the product of the characteristic polynomials of the blocks C_k, and we have

$$\det(Q - \lambda I) = (-1)^n (\lambda^{n_1} - 1)(\lambda^{n_2} - 1) \cdots (\lambda^{n_r} - 1),$$

where $n_1 + n_2 + \cdots + n_r = n = \partial u(x)$. In particular, $\det(Q) = \pm 1$.

4.8.4 The powers of matrix Q

Let us consider the following linear transformation of the algebra A:

$$\varphi^i : f(x) \to f(x)^{p^i}.$$

We shall see that the subspace of fixed points of φ^i allows us to obtain information about the factorisation type. As observed in the previous section, we may limit ourselves to study φ^i on each of the components A_k, i.e. on a field.

Theorem 4.23. *Let F_q be a field, $q = p^n$. Then the set of fixed points of the mapping $\varphi^i : \alpha \to \alpha^{p^i}$ is a subfield of F_q of order $p^{\gcd(i,n)}$.*

Proof. It is straightforward to check that the set of fixed points is closed w.r.t. sum and product. If $\alpha^{p^i} = \alpha$ and β is the inverse of α, then from $\alpha\beta = 1$ we have $1 = \alpha^{p^i}\beta^{p^i} = \alpha\beta^{p^i}$, and the uniqueness of the inverse implies $\beta^{p^i} = \beta$. Now let K be the subfield of the fixed points of φ^i and let $|K| = p^t$; let us show that $t = \gcd(i, n)$. Let α be a primitive element of K; then $\alpha^{p^t - 1} = 1$ where $p^t - 1$ is the least exponent such that the power is 1. But we also have

$\alpha^{p^i-1} = 1$, so that $p^t - 1$ divides $p^i - 1$ and so $t|i$ (see Exercise 9). On the other hand, K is a subfield of F_q, which has order p^n, and so $t|n$; it follows $t|\gcd(i,n)$. We know that F_q contains one and only one subfield L of order $p^{\gcd(i,n)}$, and L only one subfield of order p^t. K being the unique subfield of F_q of order p^t, we have $K \subseteq L$. Now, if $\beta \in L$, then $\beta^{p^{\gcd(i,n)}-1} = 1$, but since $p^{\gcd(i,n)} - 1$ divides $p^i - 1$ (because $\gcd(i,n)$ divides i) we also have $\beta^{p^i-1} = 1$, i.e. $\beta^{p^i} = \beta$ and $\beta = K$. It follows $K = L$, as required. $\diamond$

On each component A_k, the subfield of the fixed points of φ^i has order $p^{\gcd(i,n_k)}$ (notation as in the previous section), and therefore, as a vector space over Z_p, it has dimension $\gcd(i,n_k)$. Summing over all the components A_k we have the following theorem.

Theorem 4.24. *The subspace of the fixed points of the linear transformation φ^i on A has dimension $d = \sum_{k=1}^{r} \gcd(i, n_k)$.*

The matrix of the transformation φ^i is Q^i, and d is the dimension of the kernel of $Q^i - I$.

Corollary 4.15. *For $i = 2$ we have $d = r + d'$, where r is the number of irreducible factors of $u(x)$, and d' that of irreducible factors of even degree.*

Proof. Let $n_1, n_2, \ldots, n_s$ be the degrees of the irreducible factors of odd degree, and $n_{s+1}, n_{s+2}, \ldots, n_r$ those of the factors of even degree. Then

$$\gcd(i, n_k) = 1, k = 1, 2, \ldots, s,$$
$$\gcd(i, n_k) = 2, k = s + 1, s + 2, \ldots, r.$$

It follows $d = \sum_{k=1}^{r} \gcd(i, n_k) = s + 2(r - s) = 2r - s = r + (r - s)$, and $r - s$ is the number of irreducible factors whose degree n_k is even. $\diamond$

Example. Let us go back to Example 1 after Corollary 4.13. With the matrix Q of that example we have $Q^2 = I$, so that $Q^2 - I = 0$, and the kernel is the whole space. Hence $d = 4$, and since $r = 2$ we have $d' = 2$; actually, $u(x)$ has two irreducible factors of even degree (=2).

4.9 Hensel's lemma

In this section we ask ourselves whether it is possible, starting from a factorisation mod p of an integer polynomial, to obtain a factorisation mod p^k, for all k. We shall see that the answer is positive.

We have already partially solved this problem when we have considered the p-adic expansion of an algebraic number. For example, 3 is a square root of 2 mod 7, that is, a root of $u(x) = x^2 - 2$; expanding it we obtain the approximate values 3, 10, 108 etc. mod 7, 7^2, 7^3 etc., respectively. Similarly, for the other root 4 we obtain 4, 39, 235 etc., mod 7, 7^2, 7^3 etc. These expansions yield the factorisations

$$u(x) \equiv (x-3)(x+3) \bmod\ 7 \equiv (x-10)(x+10) \bmod\ 7^2$$
$$\equiv (x-108)(x+108) \bmod\ 7^3 \dots$$

taking into account that $4 \equiv -3$, $39 \equiv -10$, $235 \equiv -108$ etc. In this way, the factorisation mod 7 (indeed, the existence of a root is equivalent to a factorisation) lifts to one mod 7^k, for all k.

The following lemma yields a method to construct the factorisations mod p^k which, as suggested by the above example, is similar to the one affording the p-adic expansion of a number: once e_n is known (notation of § 2.2), that is, once the approximation mod p^n is known, the approximation mod p^{n+1} is found in the form $e_{n+1} = e_n + cp^n$.

Lemma 4.4 (Hensel's lemma). *Suppose that for a given $k \geq 1$ and a prime p we have*

$$u(x) \equiv f(x)g(x) \bmod p^k, {}^3$$

where $u(x)$, $f(x)$ and $g(x)$ have integer coefficients, are monic, and $f(x)$ and $g(x)$ are relatively prime mod p. Then there exist and are unique two polynomials $f_1(x)$ and $g_1(x)$, also monic, of degrees $\partial f_1 = \partial f$, $\partial g_1 = \partial g$ and such that $u(x) \equiv f_1(x)g_1(x) \bmod p^{k+1}$, and moreover:

$$f_1(x) \equiv f(x) \bmod p^k, \quad g_1(x) \equiv g(x) \bmod p^k.$$

Proof. (We write f for $f(x)$.) By assumption, the coefficients of $u - fg$ are all divisible by p^k, so that $u - fg = p^k c$, for some $c \in Z[x]$. Let us now express c in terms of f and g as follows. The polynomials f and g being relatively prime mod p, there exist a and b such that $af + bg \equiv 1 \bmod p$; it follows $caf + cbg \equiv c \bmod p$. Now, u, f and g are monic, and so is fg. Hence in $u - fg$ the leading term vanishes, and therefore $\partial(u - fg) < \partial u = \partial fg$, from which $\partial c < \partial fg$. From Lemma 1.1 it follows the existence of v and w such that $wf + vg \equiv c \bmod p$, i.e. $c = wf + vg + pd$, with $\partial v < \partial f$ and $\partial w < \partial g$, where v and w are uniquely determined. It follows

$$u = fg + p^k c = fg + p^k(wf + vg) + p^{k+1}d.$$

But $fg + p^k(wf + vg) = (f + p^k v)(g + p^k w) - p^{2k} vw$, and setting $f_1 = f + p^k v$, $g_1 = g + p^k w$, we have $u = f_1 g_1 - p^{2k} vw + p^{k+1}d$, and being $2k \geq k + 1$, since $k \geq 1$, we have $u \equiv f_1 g_1 \bmod p^{k+1}$.

---

[3] This congruence mod p^k means that the coefficients of the polynomial $u(x) - f(x)g(x)$ are all divisible by p^k, so that by dividing them by p^k we have a polynomial with integer coefficients.

These polynomials f_1 and g_1 fulfil the requirements of the lemma. Indeed,

1. The pair f_1, g_1 is uniquely determined by the pair f, g because this is the case for the pair v, w.
2. Since $\partial v < \partial f$, the leading monomial of f_1 equals that of f. In particular, $\partial f_1 = \partial f$, and f_1 is monic. Similarly for g_1.
3. f_1 and g_1 are relatively prime mod p: with a and b as above we have
$$af_1 + bg_1 = af + bg + (av + bw)p^k \equiv 1 \bmod p^k.$$

The lemma is proved. $\Diamond$

Remarks. 1. The gcd of two polynomials being defined up to a constant, it is possible that the calculation of the Bézout coefficients gives a value d of $af + bg$ different from 1. However, in order that the algorithm works the gcd must be 1; hence before proceeding it is necessary to divide a and b by d.

2. From 3. of the lemma it follows that, starting from a factorisation $u \equiv fg \bmod p$, the coefficients a and b such that $af + bg \equiv 1 \bmod p$ are the same as those for the pair $f^{(k)}, g^{(k)}$ such that $u \equiv f^{(k)}g^{(k)} \bmod p^k$, for all $k \geq 1$.

The following algorithm yields the factorisation mod p^{n+1}, $n = 1, 2, \ldots$, given the factorisation mod p. Notation is as in the lemma.

input: u, f, g, a, b, p, n;
output: f, g;
for k from 1 to n do:
$c := \text{quotient}(u - fg, p^k) \bmod p$;
$q := \text{quotient}(cb, f) \bmod p$;
$v := \text{remainder}(cb, f) \bmod p$;
$w := ca + qg \bmod p$;
$f := f + vp^k \bmod p^{k+1}$;
$g := g + wp^k \bmod p^{k+1}$.

Remark. Since $s \cdot p^k \bmod p^{k+1} = (s \bmod p) \cdot p^k$, for every integer s, the values v and w should be taken mod p before multiplying by p^k.

Let us now see an example of how this algorithm works.

Example. Consider the polynomial $u = x^4 - 2x^2 + 9$, with $p = 5$, $n = 4$. Let us factor u modulo 5 (using Berlekamp, for instance); we find

$$u \equiv (x^2 + x + 2)(x^2 - x + 2) \bmod 5.$$

These two factors are the input data f and g. Moreover,

$$a = x - 1 \quad \text{and} \quad b = -x - 1.$$

With these data we have

- $k = 1$:

$$c = \frac{u - fg}{5} = \frac{-5x^2 + 5}{5} = 1 - x^2.$$

Now, $cb = x^3 + x^2 - x - 1$; by dividing by f we obtain a quotient x and a remainder $2x - 1$:

$$q = x, \ v = 2x - 1, \ w = (1 - x^2)(x - 1) + x(x^2 - x + 2) = 3x - 1,$$

and $f_1 = f + 10x - 5 \equiv x^2 + 11x - 3 \bmod 5^2$, $g_1 = g + 15x - 5 \equiv x^2 - 11x - 3 \bmod 5^2$;

- $k = 2$:

with the new values of f and g we have $fg = x^4 - 127x^2 + 9$:

$$c = \frac{u - fg}{5^2} = \frac{125x^2}{5^2} = 5x^2 \equiv 0 \bmod 5.$$

In this case $v = w = 0$ so that the values of f and g are the previous ones;

- $k = 3$:

$u - fg = 125x^2$, $c = \frac{125x^2}{5^3} = x^2$, $cb = x^2(-x - 1) = -x^3 - x^2$ which divided by f gives a quotient $-x$ and a remainder $-3x$. Then $v = -3x$, $w = x^2(x - 1) - x(x^2 - 11x - 3) = 10x^2 + 3x \equiv 3x \bmod 5$, and

$$f = x^2 + 11x - 3 - 3 \cdot 125x = x^2 - 364x - 3 \equiv x^2 + 261x - 3 \bmod 5^4,$$

$$g = x^2 - 11x - 3 + 3 \cdot 125x = x^2 + 364x - 3 \equiv x^2 - 261x - 3 \bmod 5^4.$$

4.9.1 Hensel's lemma for more than two factors

The method for lifting a factorisation mod p^k to one mod p^{k+1} seen in the proof of Hensel's lemma can be extended to the case of more than two factors, as we shall now see.

Lemma 4.5 (Hensel's lemma, the general case). *Let*

$$u, u_1, u_2, \ldots, u_n \tag{4.21}$$

be monic integer polynomials such that

$$\gcd(u_i, u_j) \equiv 1 \bmod p, \ i \neq j, \tag{4.22}$$

and moreover

$$u \equiv u_1 u_2 \cdots u_n \bmod p^k. \tag{4.23}$$

Then there exist monic integer polynomials $\bar{u}_1, \bar{u}_2, \ldots, \bar{u}_n$ such that

$$\partial \bar{u}_i = \partial u_i, \ \bar{u}_i \equiv u_i \bmod p^k, \ \gcd(\bar{u}_i, \bar{u}_j) \equiv 1 \bmod p,$$

and moreover

$$u \equiv \bar{u}_1 \bar{u}_2 \cdots \bar{u}_n \bmod p^{k+1}. \tag{4.24}$$

Proof. As in the case of two factors let us seek the $\bar{u}_i$s in the form

$$\bar{u}_i = u_i + p^k v_i, \ i = 1, 2, \ldots, r, \tag{4.25}$$

for suitable v_is to be determined. Substituting the values (4.25) in (4.24) we obtain

$$u \equiv u_1 u_2 \cdots u_n + p^k (u_1 v_2 u_3 \cdots u_n + v_1 u_2 u_3 \cdots u_n + \cdots + u_1 u_2 u_3 \cdots v_n)$$

mod p^{k+1}. By (4.23),

$$c = \frac{u - u_1 u_2 \cdots u_n}{p^k}$$

is an integer polynomial, and is of smaller degree than u, i.e. the degree of $u_1 u_2 \cdots u_n$. The latter is a monic polynomial, because such are the u_i, so that subtracting u the leading monomial vanishes. Hence we must solve for the v_is the congruence

$$c \equiv u_1 v_2 u_3 \cdots u_n + v_1 u_2 u_3 \cdots u_n + \cdots + u_1 u_2 u_3 \cdots v_n \bmod p. \tag{4.26}$$

By assumption (4.22) there exist polynomials $a_1, a_2, \ldots, a_n$ such that

$$a_1 u_2 u_3 \cdots u_n + a_2 u_1 u_3 \cdots u_n + \cdots + a_n u_1 u_2 \cdots u_{n-1} \equiv 1 \bmod p. \tag{4.27}$$

Multiplying this congruence by c and dividing ca_i by u_i, $ca_i = q_i u_i + r_i$, $0 \le \partial r_i < \partial u_i$, substituting and collecting terms we find

$$u_1 u_2 \cdots u_n (q_1 + q_2 + \cdots + q_n) + r_1 u_2 u_3 \cdots u_n + r_2 u_1 u_3 \cdots u_n + \cdots$$
$$+ r_n u_1 u_2 \cdots u_{n-1} \equiv c \bmod p.$$

Now, the summands $r_i u_1 u_2 \cdots u_{i-1} u_{i+1} \cdots u_n$ and the polynomial c are of degree less than that of $u_1 u_2 \cdots u_n$, so that if $q_1 + q_2 + \cdots + q_n$ is not the null polynomial we would have a contradiction. Setting $v_i = r_i$ we have the required solution[4]. Indeed, as in the case of two factors:

1. The n-tuple of the $\bar{u}_i$ is uniquely determined by that of the u_i because this is the case for the n-tuple of the $v_i = r_i$; if there were another solution r'_i, by subtracting from the previous one we would have

$$\sum_{i=1}^{n} \left((r_i - r'_i) \prod_{j \neq i} u_j \right) \equiv 0 \bmod p.$$

---

[4] Note that the r_i are the polynomials that give the partial fraction decomposition:

$$\frac{c}{u} = \frac{r_1}{u_1} + \frac{r_2}{u_2} + \cdots + \frac{r_n}{u_n}.$$

Now, u_j divides all the products $(r_i - r_i')u_1u_2\cdots u_{i-1}u_{i+1}\cdots u_n$ for $i \neq j$, and therefore also $(r_j - r_j')u_1u_2\cdots u_{j-1}u_{j+1}\cdots u_n$, and being coprime with the u_i, $i \neq j$, it must divide $r_j' - r_j$. The latter being of degree less than u_j, it must be $r_j' - r_j = 0$, or $r_j' = r_j$, and this for all $j = 1, 2, \ldots, n$.

2. $\partial\bar{u}_i = \partial u_i$ since $\partial v_i < \partial u_i$. This inequality also implies that the leading monomial of $\bar{u}_i$ is that of u_i, and therefore $\bar{u}_i$ is monic.

3. $\gcd(\bar{u}_i, \bar{u}_j) = 1$: if $au_i + bu_j \equiv 1 \bmod p$, then $a\bar{u}_i + b\bar{u}_j = au_i + bu_j + p^k(av_i + bv_j) \equiv 1 \bmod p$. $\diamond$

Example. Let $u = x^4 - x^3 - 2x + 1$. Factoring $u \bmod 5$ (for instance using Berlekamp) we have

$$u(x) \equiv (x - 2)(x + 1)(x^2 + 2) \bmod 5.$$

Now, $u - u_1u_2u_3 = 5$, so $c = \frac{5}{5} = 1$. Solving

$$a_1(x + 1)(x^2 + 2) + a_2(x - 2)(x^2 + 2) + a_3(x - 2)(x + 1) \equiv 1 \bmod p,$$

we find $a_1 = 2$, $a_2 = 1$, $a_3 = 2x + 2$, and since $\partial ca_i = \partial a_i < \partial u_i$, we have directly, with no need to divide by u_i, that $v_i = a_i$ and therefore:

$$\bar{u}_1 = u_1 + 5v_1 = x - 2 + 5 \cdot 2 = x + 8,$$
$$\bar{u}_2 = u_2 + 5v_2 = x + 1 + 5 \cdot 1 = x + 6,$$
$$\bar{u}_3 = u_3 + 5v_3 = x^2 + 2 + 5(2x + 2) = x^2 + 10x + 12.$$

From this follows

$$x^4 - x^3 - 2x + 1 = (x + 8)(x + 6)(x^2 + 10x + 12) \bmod 25.$$

4.10 Factorisation over the integers

We may now use the results of the two previous sections in order to factor integer polynomials. We begin with the following observation.

Let $u(x) = F(x)G(x)$ be a factorisation over Z of $u(x)$. Modulo an integer M we have $u(x) \equiv f(x)g(x) \bmod M$, with

$$F(x) \equiv f(x) \bmod M, \quad G(x) \equiv g(x) \bmod M. \tag{4.28}$$

Suppose we know that the coefficients of every possible factor of $u(x)$ are not greater, in modulus, than a given integer B (in the next section we will see how to find such an integer). Then, if $M > 2B$, if we choose the coefficients of $f(x)$ in the interval $[-M/2, M/2]$ (i.e., if we choose the balanced representation of the integers mod M), relations (4.28) are not only congruences, but actually equalities. Indeed, if $F(x) \neq f(x)$, there exists a coefficient a of

$F(x)$ differing from the coefficient b of the monomial of $f(x)$ having the same degree by a multiple of M: $a - b = hM$. Therefore $a = b + hM$, and with the possible values for b we have $|a| \geq M/2 > B$, contrary to assumption.

Suppose now that for a prime $p > 2B$ a given polynomial splits mod p into the product of two factors:

$$u(x) \equiv f(x)g(x) \bmod p. \qquad (4.29)$$

Then we may know whether this factorisation comes from a factorisation

$$u(x) = F(x)G(x) \qquad (4.30)$$

over the integers. Indeed, choosing the coefficients of $f(x)$ between $-(p-1)/2$ and $(p-1)/2$, if (4.29) comes from (4.30) we have $f(x) = F(x)$, and therefore $f(x)$ must divide $u(x)$ over Z.

In order to find factors of $u(x)$ we may then proceed as follows.

1. Factor $u(x) \bmod p$, with $p > 2B$, into irreducible factors, for instance using Berlekamp's method.

2. For each factor $p_i(x)$, considered as a polynomial with integer coefficients between $[-(p-1)/2, (p-1)/2]$, check if it divides $u(x)$ over Z. A $p_i(x)$ that divides $u(x)$ is an irreducible factor of $u(x)$ (if it reduces, it also reduces mod p).

3. If there are $p_i(x)$s that do not divide $u(x)$, form with these the products $p_i(x)p_j(x)$, reduce the coefficients to the mentioned interval, and see whether these product polynomials divide $u(x)$. Every polynomial constructed in this way that divides $u(x)$ is an irreducible factor of $u(x)$ (if it reduces, it necessarily reduces into the product of $p_i(x)$ and $p_j(x)$, which then would divide $u(x)$); cancel the $p_i(x)$s that are part of a factor of $u(x)$.

4. With the remaining $p_i(x)$ form the products $p_i(x)p_j(x)p_k(x)$ and, again after reducing the coefficients to the interval $[-(p-1)/2, (p-1)/2]$, check if they divide $u(x)$.

5. Continue until all the combinations have been tried.

6. If some $p_i(x)$s remain, their product is an irreducible factor of $u(x)$.

Instead of considering a prime $p > 2B$ (B may be very large), one can take any prime p, factor mod p, and then lift the factorisation to p^n, where n is such that $p^n > 2B$.

4.10.1 Upper bounds for the coefficients of a factor

Let us see now how to determine the integer B of the previous section. That such an integer exists depends on the fact that it is possible to give an upper bound for the modulus of the roots of a polynomial, as the following theorem shows.

Theorem 4.25. *Let*

$$u(x) = a_0 x^n + a_1 x^{n-1} + \cdots + a_n$$

be a polynomial with complex coefficients. Then for any root z of $u(x)$ we have:

$$|z| < 1 + \frac{M}{|a_0|}, \tag{4.31}$$

where $M = \max\{|a_0|, |a_1|, \ldots, |a_n|\}$.

Proof. If $|z| \leq 1$ there is nothing to prove. Assume $|z| > 1$. Since $u(z) = 0$ we have

$$a_0 z^n = -a_1 z^{n-1} - a_2 z^{n-2} - \cdots - a_n.$$

It follows:

$$|a_0||z|^n \leq |a_1||z|^{n-1} + |a_2||z|^{n-2} + \cdots + |a_n| \leq M(|z|^{n-1} + |z|^{n-2} + \cdots + 1)$$
$$= M \frac{|z|^n - 1}{|z| - 1} < M \frac{|z|^n}{|z| - 1},$$

from which $|z| < 1 + \frac{M}{|a_0|}$, as required. $\diamond$

Let now $f(x)$ be a monic polynomial with roots z_i:

$$f(x) = x^m + b_1 x^{m-1} + \cdots + b_m = (x - z_1)(x - z_2) \cdots + (x - z_m).$$

The coefficients b_i are the elementary symmetric functions of the z_i:

$$b_1 = -(z_1 + z_2 + \cdots + z_m),\ b_2 = z_1 z_2 + z_1 z_3 + \cdots + z_{m-1} z_m, \ldots,$$
$$b_k = (-1)^k \sum z_{i_1} z_{i_2} \cdots z_{i_k}, \ldots, b_m = (-1)^m z_1 z_2 \cdots z_m.$$

If $|z_i| \leq B$, then

$$|b_1| = |\sum z_i| \leq \sum |z_i| \leq mB,$$
$$|b_2| = |\sum_{i,j} z_i z_j| \leq \sum_{i,j} |z_i||z_j| \leq \binom{m}{2} B^2,$$
$$\vdots$$
$$|b_m| = |z_1||z_2| \cdots |z_m| \leq B^m,$$

from which $|b_i| \leq \max\{\binom{m}{k} B^k,\ k = 1, 2, \ldots, m\}$.

If $f(x)$ is a factor of a polynomial $u(x)$, a root of $f(x)$ is also a root of $u(x)$: an upper bound for the coefficients of $f(x)$ can then be obtained from one for the roots of $u(x)$ (for example, the one given by (4.31)).

Another bound (that we will not prove) for the modulus of the coefficients of a factor may be obtained as follows. Given the polynomial

$$u(x) = u_n x^n + u_{n-1} x^{n-1} + \cdots + u_0,$$

define

$$\|u\| = (|u_n|^2 + |u_{n-1}|^2 + \cdots + |u_0|^2)^{\frac{1}{2}}.$$

Let $f(x) = b_m x^m + b_{m-1} x^{m-1} + \cdots + b_0$ be a factor of $u(x)$. Then

$$|b_j| \leq \binom{m-1}{j} \|u\| + \binom{m-1}{j-1} |u_n|.$$

Example. Let $u(x) = x^8 + x^6 - 3x^4 - 3x^3 + 8x^2 + 2x - 5$. If $u(x)$ is reducible, then it must have a factor of degree at most 4. Let $f(x) = b_4 x^4 + b_3 x^3 + b_2 x^2 + b_1 x + b_0$. We have $\|u\| = \sqrt{113} \sim 10.6$; hence:

$$|b_0| \leq \binom{3}{0} \|u\| + 0 \sim 10.6; \ |b_1| \leq \binom{3}{1} \|u\| + \binom{3}{0} \sim 31.9;$$

$$|b_2| \leq \binom{3}{2} \|u\| + \binom{3}{1} \sim 34.9; \ |b_3| \leq \binom{3}{3} \|u\| + \binom{3}{2} \sim 13.6.$$

It follows that no coefficient of $f(x)$ is greater than 34 in modulus.

We can now see whether $u(x)$ factors over Z using the method seen above. Choose a prime $p > 2 \cdot 34 = 68$, for instance $p = 71$. Berlekamp's method yields

$$u(x) \equiv (x + 12)(x + 25)(x^2 - 13x - 7)(x^4 - 24x^3 - 16x^2 + 31x - 12) \bmod 71.$$

None of these factors divides $u(x)$ over Z (because none of the constant terms divides 5). By grouping the factors in pairs we have the constant terms $12 \cdot 25 = 300$, $12 \cdot (-7) = -84$, $12 \cdot (-12) = -144$, none of which is congruent to ± 1 nor to $\pm 5 \bmod 71$. We conclude that $u(x)$ is irreducible over the integers.

The procedure to factor over Z we have seen (Berlekamp's method possibly followed by a Hensel lifting) has the drawback of being of exponential algorithmic complexity[5] with respect to the degree n of the polynomial $u(x)$. When the factors of the polynomial mod p are all linear and $u(x)$ is irreducible over Z, before realising that the polynomial is irreducible one must perform 2^n divisions, i.e. as many divisions as the total number of one by one, two by two, etc., combinations of the n factors. However, there exists a method called "L^3"[6], which allows the factorisation in polynomial time.

---

[5] For a brief discussion about algorithmic complexity see next chapter.

[6] From the initials of A.K. Lenstra, H.W. Lenstra and L. Lovász, authors of the paper [**LLL**].

4.11 Factorisations in an extension

Let K be the splitting field of the monic irreducible polynomial $p(x)$ of degree m with coefficients in the rational field Q, and let $\alpha_1, \alpha_2, \ldots, \alpha_m$ be its roots (which are all distinct because $p(x)$ is irreducible and the characteristic of Q is zero). Let us consider the subfield $Q(\alpha)$ of K consisting of the polynomials in α over Q of degree at most $n-1$. If $\beta \in Q(\alpha)$, then $\beta = a_0 + a_1\alpha + \cdots + a_{m-1}\alpha^{m-1}$ with $a_i \in Q$. The *norm* of β is defined as

$$N(\beta) = \prod_{i=1}^{m}(a_0 + a_1\alpha_i + \cdots + a_{m-1}\alpha_i^{m-1}).$$

If $g(x)$ is the polynomial $a_0 + a_1x + \cdots + a_{m-1}x^{m-1}$, then

$$N(\beta) = \prod_{p(x)=0} g(x) = R(p, g),$$

which is the resultant of the polynomials p and g. In particular, the norm of an element of $Q(\alpha)$ belongs to Q, i.e. is a rational number. Now let $f = f(x)$ be a polynomial with coefficients in $Q(\alpha)$, $f(x) \in Q(\alpha)[x]$. This may be considered as a polynomial $f(\alpha, x)$ in the two variables α and x, with coefficients in Q. We then define the *norm of the polynomial f* as

$$N(f) = \prod_{i=1}^{n} f(\alpha_i, x) = \prod_{p(y)=0} f(y, x) = R_y(p(y), f(y, x)).$$

Then the norm of $f \in Q(\alpha)[x]$ is a rational polynomial, $N(f) \in Q[x]$, of degree $\partial f \cdot \partial p$, the product of the degrees of f and of the polynomial p of which α is a root. If $f \in Q[x]$, then $N(f) = f$ (in this case α is just one of the coefficients of f; this can also be seen from the resultant because now $p(y) = y - \alpha$, and $R_y(y-\alpha, f(y,x)) = f(\alpha, x) = f(x)$). Moreover, by a property of the resultant, the norm is multiplicative: $N(fg) = N(f)N(g)$.

We now prove that the factorisation of a rational polynomial reduces to that of its norm.

Remark. Let $(p(x)$ be an irreducible polynomial over a field F and let K be its splitting field. The *Galois group* of K over F, denoted $G(K/F)$, is the group of automorphisms of K leaving F elementwise fixed. K being the splitting field of a polynomial, there are no other elements of K fixed by elements of $G(K/F)$. If $\sigma \in G(K/F)$, then σ extends in a natural way to the polynomials over K: if $f(x) = a_0 + a_1x + \cdots + a_nx^n$, then $\bar{\sigma}(f(x)) = \sigma(a_0) + \sigma(a_1)x + \cdots + \sigma(a_n)x^n$. In particular, if $f(x)$ has coefficients in F, then $\bar{\sigma}(f(x)) = f(x)$. If α is a root of our polynomial $p(x) \in F[x]$, the element $\sigma(\alpha)$, $\sigma \in G(K/F)$, is again a root of $p(x)$. Hence, the Galois group induces a permutation group of the roots and, K being the splitting field of the irreducible polynomial $p(x)$, it can be shown that, given two roots of $p(x)$, there is an element of $G(K/F)$ taking one to the other.

Lemma 4.6. *Let $f \in Q(\alpha)[x]$ be irreducible. Then $N(f)$ is a power of an irreducible polynomial.*

Proof. Let $N(f(x)) = h(x)g(x)$, with $h(x), g(x) \in Q[x]$ and $\gcd(h(x), g(x)) = 1$. By definition, $N(f(x)) = \prod_{i=1}^{n} f(\alpha_i, x)$, and therefore $f(\alpha, x)$ must divide either h or g in $Q(\alpha)[x]$. Let $h(x) = f(\alpha, x)h_1(\alpha, x)$. For every α_i there exists $\sigma_i \in G(K/Q)$ such that $\sigma(\alpha) = \alpha_i$, and therefore

$$\bar{\sigma}(h(x)) = h(x) = \bar{\sigma}(f(\alpha, x))\bar{\sigma}(h_1(\alpha, x)) = f(\alpha_i, x)h_1(\alpha_i, x).$$

Hence all the polynomials $f(\alpha_i, x)$ divide h, and therefore none divides g because $\gcd(h, g) = 1$. Then $N(f) = h$, and h is either irreducible, or is a power of an irreducible polynomial.   ◇

Theorem 4.26. *Let $f \in Q(\alpha)[x]$ with $N(f)$ square-free. If*

$$N(f) = \prod_i G_i(x)$$

is the factorisation of the norm of f into irreducible factors in $Q(x)$, then

$$f = \prod_i \gcd(f(x), G_i(x))$$

is the factorisation of f into irreducible factors in $Q(\alpha)[x]$.

Proof. Let us prove first that the gcds $g_i = (f, G_i)$ are irreducible. If $h_i \in Q(\alpha)[x]$ is a non-trivial factor of g_i, then it is also a non-trivial factor of G_i, so that $G_i = h_i k_i$, for some k_i. Taking norms, $G_i = N(G_i) = N(h_i)N(k_i)$, and by the irreducibility of G_i, one of the two norms must be a constant. Hence, one of the two polynomials, and so also the other one, belongs to $Q(x)$; but this contradicts the irreducibility of G_i in $Q[x]$. Moreover, the g_i are distinct because so are the G_i.

If $f = \prod_j h_j$ is the factorisation of f into irreducible factors in $Q(\alpha)[x]$, we have

$$N(f) = \prod_j N(h_j) = \prod_i G_i.$$

By Lemma 4.6, $N(h_j)$ is a power of an irreducible polynomial, and each $N(h_j)$ must divide $N(f)$. But $N(f)$ is square-free, and therefore $N(h_j)$ is one of the irreducible factors of $N(f)$ and so equals one of the G_i. Moreover, $\gcd(N(h_i), N(h_j)) = 1$ if $i \neq j$, again because $N(f)$ is square-free. By the uniqueness of the factorisation it follows that, up to ordering, the h_j and the g_i are equal.   ◇

If $N(f)$ has multiple roots we may use the following result.

Theorem 4.27. *Let $f(x) \in Q(\alpha)[x]$ be a square-free polynomial of degree n. Then there exist at most $\frac{(nm)^2}{2} - 1$ integers s such that $N(f(x - s\alpha))$ has multiple roots.*

Proof. If $\beta_1, \beta_2, \ldots, \beta_n$ are the roots of $f(x)$, then $\beta_i + s\alpha_j$, $i = 1, 2, \ldots, n$, $j = 1, 2, \ldots, m$, are the roots of $N(f(x - s\alpha))$. The multiple roots are those for which $\beta_i + s\alpha_j = \beta_k + s\alpha_t$, with $i \neq k$ and $j \neq t$, i.e. those for which $s = \frac{\beta_k - \beta_i}{\alpha_j - \alpha_t}$. The (ordered) pairs of the β are $n(n-1)$, those of the αs are $m(m-1)$, and two pairs (β_k, β_i) and (α_j, α_t) give for s the same value of the pairs (β_i, β_k) and (α_t, α_j). Thus the integer s can assume at most $\frac{n(n-1)m(m-1)}{2} < \frac{(nm)^2}{2}$ values.
$\diamond$

Hence, proceed as follows to factor $f(x)$:

1. If f is not square-free, replace f with f/d, where $d = (f, f')$;

2. Look for an integer s such that $N(f(x - s\alpha))$ is square-free;

3. Factor $N(f(x - s\alpha))$ in $Q[x]$:

$$N(f(x - s\alpha)) = \prod_i H_i(x);$$

4. By Theorem 4.26,

$$f(x - s\alpha) = \prod_i \gcd(H_i(x)f(x - s\alpha))$$

is the complete factorisation of $f(x - s\alpha)$ in $Q(\alpha)[x]$;

5. Replacing $x \to x + s\alpha$ the factorisation of $f(x)$ is complete:

$$f(x) = \prod_i \gcd(H_i(x + s\alpha), f(x)).$$

Example. Let $K = Q(\sqrt{2})$, and let $f(x) = x^3 - (1 + \sqrt{2})x^2 + \sqrt{2}x - \sqrt{2} - 2$. The minimal polynomial of $\sqrt{2}$ is $p(y) = y^2 - 2$, and the norm of $f(x)$ is $N(f) = R_y(p(y), f(x, y)) = x^6 - 2x^5 - x^4 - 2x^2 + 4x + 2 = (x^2 - 2x - 1)(x^4 - 2)$; therefore $N(f)$ is square-free. Setting $G_1(x) = x^2 - 2x - 1$ and $G_2(x) = x^4 - 2$, we have $\gcd(f(x), G_1(x)) = x - 1 - \sqrt{2}$, $\gcd(f(x), G_2(x)) = x^2 + \sqrt{2}$, from which the factorisation $f(x) = (x - 1 - \sqrt{2})(x^2 + \sqrt{2})$ follows.

References

Finite fields and Berlekamp's method are dealt with in various books, in particular [C] II-12, [McE] Chapters 6 and 7, [LN] Chapter IV, [DST], section 4.2, and [Kn], p. 420ff. In [DST] Hensel's lemma is also discussed, as well as in [Kn], p. 439, ex. 22. For section 4.5 we have followed [DST], section 1. For the upper bound of the coefficients of a factor, [CA] p. 259 and [Kn], p. 438, ex. 20.

5

The discrete Fourier transform

5.1 Roots of unity

The n-th roots of unity are the roots of the polynomial $x^n - 1$ in the complex field. We know that they are all distinct because the polynomial is coprime with its derivative, and that they are all powers of one of them, a primitive root $w = e^{2\pi i/n}$:

$$1, w, w^2, \ldots, w^{n-1},$$

with $w^n = 1$. (Recall that the primitive roots w^k are those for which $(k, n) = 1$, so that their number is $\varphi(n)$, where φ is Euler's function.) Since 1 is a root of $x^n - 1$, this polynomial is divisible by $x - 1$, with quotient $1 + x + x^2 + \cdots + x^{n-1}$, and therefore all the w^k, with $k \neq 0$ (or, actually, $k \not\equiv 0 \bmod n$), satisfy the equation:

$$1 + x + x^2 + \cdots + x^{n-1} = 0. \tag{5.1}$$

In particular, for $x = w$ we see that the sum of all the n-th roots of unity is zero, $1 + w + w^2 + \cdots + w^{n-1} = 0$. Moreover, since $|w| = 1$, we have $|w|^k = 1$, so that $w^k \overline{w}^k = |w^k|^2 = 1$, from which

$$\overline{w}^k = \frac{1}{w^k} = w^{-k}. \tag{5.2}$$

We use formulas (5.1) and (5.2) to prove the following theorem; despite its simplicity, it will be of fundamental importance for the entire discussion.

Theorem 5.1. *We have:*

$$\sum_{j=0}^{n-1} w^{hj} \overline{w}^{kj} = \begin{cases} n, \ \textit{if } h \equiv k \bmod \ n; \\ 0, \ \textit{otherwise.} \end{cases} \tag{5.3}$$

Machì A.: Algebra for Symbolic Computation.
DOI 10.1007/978-88-470-2397-0_5, © Springer-Verlag Italia 2012

Proof. Because of (5.2), the sum of (5.3) in extended form is

$$1 + w^{h-k} + w^{2(h-k)} + \cdots + w^{(n-1)(h-k)}.$$

If $h \equiv k \bmod n$, this is the sum of n times 1, and therefore it equals n. Otherwise, w^{h-k} is an n-th root of unity different from 1, so it satisfies (5.1); hence the sum equals zero. $\diamond$

Relations (5.3) are called *orthogonality relations*.

5.1.1 Interpolation at the roots of unity

A polynomial $u(x) = u_0 + u_1 x + \cdots + u_{n-1} x^{n-1}$ of degree less than n is uniquely determined when the values $z_0, z_1, \ldots, z_{n-1}$ it assumes at n distinct points $x_0, x_1, \ldots, x_{n-1}$ are known. Hence the polynomial may be represented in two ways:

i) through the n-tuple of its coefficients:

$$u = (u_0, u_1, \ldots, u_{n-1});$$

ii) through the n-tuple of the values assumed at the n points x_i:

$$z = (z_0, z_1, \ldots, z_{n-1}).$$

If the z_is are known, one can go back to the u_is by solving the system of equations

$$u_0 + u_1 x_i + u_2 x_i^2 \cdots + u_{n-1} x_i^{n-1} = z_i, \tag{5.4}$$

in the unknowns u_i, $i = 0, 1, \ldots, n-1$, which admits exactly one solution because its determinant is the Vandermonde of the x_is, and so is non zero because the x_is are distinct. Hence the solution may be found by Cramer's rule. In case the x_is are the roots of unity the system may be solved in a different way, by making use of the orthogonality relations (5.3). Consider the z_is as coefficients of a new polynomial $v(x) = z_0 + z_1 x + \cdots + z_{n-1} x^{n-1}$; then the k-th unknown u_k is the value of the polynomial $\frac{1}{n} v(x)$ at the point $\overline{w}^k$, as the following theorem shows.

Theorem 5.2. *Let $u(x)$ be a polynomial whose values at the points w^i, $i = 0, 1, \ldots, n-1$, are known. Then the coefficients $u_0, u_1, \ldots, u_{n-1}$ of $u(x)$ are given by:*

$$u_k = \frac{1}{n}(z_0 + z_1 \overline{w}^k + z_2 \overline{w}^{2k} + \cdots + z_{n-1} \overline{w}^{(n-1)k}), \tag{5.5}$$

$k = 0, 1, \ldots, n-1$.

Proof. Consider the system (5.4) with $x_i = w^i$:

$$u_0 + u_1 w^i + u_2 w^{2i} + \cdots + u_{n-1} w^{(n-1)i} = z_i; \tag{5.6}$$

multiply equation number 0 by $1 = \overline{w}^0$, equation number 1 by $\overline{w}^k$, equation number 2 by $\overline{w}^{2k}$ etc., and sum the equations thus obtained; we obtain:

$$\sum_{i=0}^{n-1} z_i \overline{w}^{ik} = \sum_{i=0}^{n-1}\sum_{h=0}^{n-1} u_h w^{ih}\overline{w}^{ik} = \sum_{h=0}^{n-1} u_h \sum_{i=0}^{n-1} w^{ih}\overline{w}^{ik}.$$

By the orthogonality relations (5.3) the last sum equals zero for $h \neq k$. Only the term $u_k \sum_{i=0}^{n-1} w^{ik}\overline{w}^{ik}$ remains; in it, again by (5.3), the sum equals n. The result follows. $\diamond$

Remark. Note that the constant term u_0 of $u(x)$ is the average of the values z_i.

The n-tuple u given by (5.5) is the *Fourier transform* of the n-tuple z, and the coefficients u_k are the *Fourier coefficients* of z. In order to stress the difference with the classical case, u is called the Discrete Fourier Transform (DFT) of z.

The matrix of the system (5.6) is

$$F = F_n = \begin{pmatrix} 1 & 1 & 1 & \cdots & 1 \\ 1 & w & w^2 & \cdots & w^{n-1} \\ 1 & w^2 & w^4 & \cdots & w^{2(n-1)} \\ \vdots & \vdots & \vdots & \ddots & \vdots \\ 1 & w^{n-1} & w^{2(n-1)} & \cdots & w^{(n-1)(n-1)} \end{pmatrix},$$

the *Fourier matrix*. It is the Vandermonde matrix of the n-th roots of unity, which in this case is symmetric. By (5.3), the columns v_k, $k = 0, 1, \ldots, n-1$, are vectors (of the space C^n of the n-tuples of complex numbers) that are orthogonal w.r.t. the hermitian scalar product $(x, y) = \sum_{i=0}^{n-1} x_i \overline{y}_i$ (they are orthonormal if one divides by $\sqrt{n}$). Moreover, they are independent (both because of orthogonality and because the matrix is Vandermonde). For the u_ks of (5.5) we may then write

$$u_k = \frac{(z, v_k)}{(v_k, v_k)},$$

so $u_k v_k$ is the projection of the vector z on the vector v_k. In the basis $\{v_k\}$, z is expressed as

$$z = u_0 + u_1 v_1 + \cdots + u_{n-1} v_{n-1}.$$

The inverse F^{-1} of F represents the passage from the z_is to the u_is, and is the matrix of the system (5.5). Hence it is sufficient to take the coefficients of F, change them into their conjugate (i.e., by (5.2), into their inverses), and then divide by n:

$$F^{-1} = \frac{1}{n}\overline{F}.$$

The DFT of the n-tuple z is then expressed as[1]:

$$F^{-1}z = u.$$

The matrix U obtained from F by dividing each element by its norm $\sqrt{n}$ is unitary[2], that is, the conjugate transpose equals the inverse: $\overline{U}^t = U^{-1}$. Indeed, by the orthogonality relations (5.3),

$$U\overline{U}^t = \frac{1}{\sqrt{n}}F \cdot \frac{1}{\sqrt{n}}\overline{F}^t = F \cdot \frac{1}{n}\overline{F} = F \cdot F^{-1} = I.$$

Again by (5.3), U^2 is the matrix:

$$\tau = \begin{pmatrix} 1 & 0 & 0 & \dots & 0 \\ 0 & 0 & 0 & \dots & 1 \\ \vdots & \vdots & \vdots & \ddots & \vdots \\ 0 & 0 & 1 & 0 & 0 \\ 0 & 1 & 0 & \dots & 0 \end{pmatrix},$$

and $U^4 = 1$ since $\tau^2 = I$. In particular, the eigenvalues of U are the fourth roots of unity $1, i, -1, -i$ with the appropriate multiplicities and therefore *the eigenvalues of F are, up to a factor $\sqrt{n}$, the fourth roots of unity.*

Remark. We recall a few notions of the classical theory of the Fourier transform. In a finite-dimensional vector space the length of a vector is defined as the square root of the norm, i.e. of the sum of the squares of the components of the vector. In the infinite-dimensional case one considers the subspace of the vectors for which the series of the squares of the components converges to a finite number (Hilbert space). By definition, the square root of this number will be the length of the vector. Now, if we consider the space of real functions defined on an interval $[a, b]$, a function of this space $f = f(x)$ may be thought of as a vector having a continuous infinity of components (the values of f at the points of the interval) and the above series is replaced by an integral. The Hilbert space will then consist of the functions for which the integral $\int_a^b f(x)^2 dx$ exists and is finite (square-summable functions). Let us now consider the interval $[0, 2\pi]$ and the functions:

$$1, \cos(kx), \sin(kx),$$

for $k = 1, 2, \dots$ For these functions,

$$\int_0^{2\pi} 1^2 dx = 2\pi, \quad \int_0^{2\pi} \cos^2(kx)dx = \int_0^{2\pi} \sin^2(kx)dx = \pi,$$

---

[1] For simplicity, here and in what follows we denote by z the column vector formed by the elements of the n-tuple z, rather than by z^t (t denoting transposition), as it should be.

[2] According to some authors, this U is the Fourier matrix; according to others the Fourier matrix is $\overline{U}$, and (5.6) the Fourier transform, i.e. the one given by F.

for all k, and therefore they are vectors of length $\sqrt{2\pi}$, $\sqrt{\pi}$ and $\sqrt{\pi}$, respectively. The integral of the square is the special case $g = f$ of the scalar product defined as:

$$(f, g) = \int_0^{2\pi} f(x)g(x)dx.$$

With respect to this product the previous functions are orthogonal, that is,

$$(1, \cos(kx)) = 0, \ (1, \sin(kx)) = 0, \ (\cos(kx), \sin(kx)) = 0,$$

because of the known properties of the integrals of the trigonometric functions.

Under suitable hypotheses, a function f that is periodic of period 2π can be expanded in a *Fourier series* in the interval $[0, 2\pi]$:

$$f(x) = \frac{a_0}{2} + a_1 \cos x + b_1 \sin x + \cdots + a_k \cos kx + b_k \sin kx + \cdots .$$

The coefficients a_k and b_k are the *Fourier coefficients* of f. In order to obtain one of these coefficients a_k (b_k) multiply both sides of the equality by the corresponding cosine (sine) and then integrate between 0 and 2π. Orthogonality implies that all the terms go to zero, except the one containing the required coefficient whose value is $a_k \int_0^{2\pi} \cos^2(kx)$; similarly for b_k. For instance,

$$b_1 = \frac{\int_0^{2\pi} f(x)\sin(x)dx}{\int_0^{2\pi} \sin^2(x)} = \frac{1}{\pi}\int_0^{2\pi} f(x)\sin(x)dx,$$

or, in terms of scalar product and more generally:

$$a_k = \frac{(f, \cos(kx))}{(\cos(kx), \cos(kx))}, \quad b_k = \frac{(f, \sin(kx))}{(\sin(kx), \sin(kx))}.$$

As in the finite case, then, the summands of the Fourier series of f are the projections of f, thought of as a vector of the Hilbert space, on the lines generated by the infinitely many vectors $1, \cos(kx)$ and $\sin(kx)$. (Note that a coefficient like a_k is the value of r minimising the difference $f(x) - r\cos(kx)$; indeed, $f(x) - a_k \cos(kx)$ is the length of the segment of the perpendicular line projecting f on the line through $\cos(kx)$.)

By Euler's equality $e^{ijx} = \cos(jx) + i\sin(jx)$, where i is the imaginary unit, the expansion in sines and cosines is then equivalent to

$$f(x) = \sum_{j=-\infty}^{+\infty} c_j e^{ijx}.$$

Indeed, from the latter equality:

$$f(x) = \sum_{j=0}^{\infty}(c_j e^{ijx} + c_{-j}e^{-ijx}) = 2c_0 + \sum_{j=1}^{\infty}(c_j + c_{-j})\cos(jx) + i(c_j - c_{-j})\sin(jx),$$

and the two expansions are equal for $a_j = c_j + c_{-j}$ and $b_j = i(c_j - c_{-j})$. The c_js are the coefficients of f in this new expansion. In the case of a complex space of finite dimension n, the scalar product is the hermitian product

$$(u, v) = \sum_{i=0}^{n} x_i \bar{y}_i$$

(x_i and y_i are the coordinates of u and v, respectively). In our case,

$$(f, g) = \int_0^{2\pi} f(x)\overline{g(x)}dx.$$

With respect to this product, the functions 1 and e^{ikx} are orthogonal:

$$\int_0^{2\pi} 1 \cdot e^{ikx}dx = 0, \quad \int_0^{2\pi} e^{ikx}e^{-ijx}dx = \begin{cases} 2\pi, & \text{if } k = j; \\ 0, & \text{if } k \neq j \end{cases}$$

(recall that $\overline{e^{ijx}} = e^{-ijx}$). These orthogonality relations, and the analogous ones for sines and cosines seen above, correspond to (5.3) of the discrete case.

In order to determine a coefficient c_k, we may proceed as already seen, by multiplying $f(x)$ and its expansion by e^{-ikx} and then integrating from 0 to 2π. Keeping in mind the orthogonality relations, we get:

$$c_k = \frac{1}{2\pi} \int_0^{2\pi} f(x)e^{-ikx}dx,$$

or:

$$c_k = \frac{(f(x), e^{ikx})}{(e^{ikx}, e^{ikx})},$$

and the summand $c_k e^{ikx}$ in the series turns out to be the projection on the e^{ikx} "axis" of the function ("vector") $f(x)$.

What we have seen for DFT is the discrete version of the usual continuous case we have just discussed. In the discrete case, the series is finite, that is, a polynomial, and the multiplication by e^{-ikx} followed by the integration from 0 to 2π becomes a multiplication by $\overline{w}^k$, that is, by w^{-k}, followed by summing from 0 to n. In other words, the discrete version is the usual approximation, using the trapezoid rule, of the integral that gives c_k when we take finitely many values of $f(x)$ at suitably chosen points. Recall that in the trapezoid rule we divide the interval $[a, b]$ into equal parts using the points $a = x_0, x_1, \ldots, x_n = b$, and the area $\int_a^b g(x)dx$ under the curve $y = g(x)$ is approximated by the sum of the areas of the trapezoids having vertices $(x_k, 0)$, $(x_k, g(x_k))$, $(x_{k+1}, g(x_{k+1}))$, $(x_{k+1}, 0)$. This sum equals:

$$\frac{b - a}{n} \left(\frac{1}{2}(g(a) + g(b)) + \sum_{i=1}^{n-1} g(x_i) \right).$$

In our case, for the integral giving c_k, we have $g(x) = f(x)e^{-ikx}$, with $a = 0$ and $b = 2\pi$. Take $x_j = \frac{2\pi j}{n}$; since $f(x)$ is periodic with period 2π, we have $f(2\pi) = f(0)$, so with $z_j = f(\frac{2\pi j}{n})$ we have the following approximate value of c_k:

$$\hat{c}_k = \frac{1}{2\pi} \cdot \frac{2\pi}{n} \left(\frac{1}{2} \cdot 2z_0 + \sum_{j=1}^{n-1} z_j e^{-ik\frac{2\pi j}{n}} \right)$$

$$= \frac{1}{n} \left(z_0 + \sum_{j=0}^{n} z_j \overline{w}^{jk} \right),$$

where $w = e^{\frac{2\pi i}{n}}$. The values $\hat{c}_k$ so obtained are exactly the u_ks in (5.5).

5.2 Convolution

Let $u = (u_0, u_1, \ldots, u_{n-1})$ and $v = (v_0, v_1, \ldots, v_{n-1})$ be two n-tuples of complex numbers. The *convolution product*, or just *convolution* of the two n-tuples, denoted by $u * v$, is the n-tuple having as coefficients

$$c_k = \sum_{i+j \equiv k \bmod n} u_i v_j, \tag{5.7}$$

$k = 0, 1, \ldots, n - 1$. In explicit form:

$$
\begin{aligned}
c_0 &= u_0 v_0 + u_1 v_{n-1} + u_2 v_{n-2} + \cdots + u_{n-1} v_1, \\
c_1 &= u_0 v_1 + u_1 v_0 + u_2 v_{n-1} + \cdots + u_{n-1} v_2, \\
&\;\;\vdots \\
c_{n-1} &= u_0 v_{n-1} + u_1 v_{n-2} + u_2 v_{n-3} + \cdots + u_{n-1} v_0,
\end{aligned}
\tag{5.8}
$$

or:

$$u * v = Mu,$$

where M is the matrix of the v_is. Equality (5.7) can also be written:

$$c_k = \sum_{i=0}^{n-1} u_i v_{k-i}, \tag{5.9}$$

taking indices modulo n. Let now $z = Fu$ and $x = Fv$; the k-th coefficients of z and of x are

$$z_k = \sum_{s=0}^{n-1} u_s w^{ks}, \quad x_k = \sum_{t=0}^{n-1} v_t w^{kt}, \tag{5.10}$$

respectively. Consider the convolution $z * x$; for the k-th coefficient of this product we have:

$$y_k = \sum_{i=0}^{n-1} z_i x_{k-i} = \sum_{i=0}^{n-1} \left(\sum_{s=0}^{n-1} u_s w^{is} \cdot \sum_{t=0}^{n-1} v_t w^{(k-i)t} \right) \tag{5.11}$$

$$= \sum_{s,t=0}^{n-1} u_s v_t w^{kt} \cdot \sum_{i=0}^{n-1} w^{is} w^{-it} = n \sum_{s=0}^{n-1} u_s v_s w^{ks},$$

where the last equality follows from the orthogonality relations. So we see that the k-th element of the n-tuple $z * x$ equals n times the k-th element of the n-tuple $F(u \cdot v)$, where the product $u \cdot v$ is the n-tuple $(u_0 v_0, u_1 v_1, \ldots, u_{n-1} v_{n-1})$, that is, the n-tuple of the termwise products. From this follows:

$$F(u) * F(v) = nF(u \cdot v). \tag{5.12}$$

From this equality we deduce a rule to compute the convolution. Indeed, from $u = F(F^{-1}u)$ and the analogue for v, we have

$$u * v = F(F^{-1}u) * F(F^{-1}v) = nF[(F^{-1}u) \cdot (F^{-1}v)], \qquad (5.13)$$

or, computing a convolution is achieved by two Fourier transforms plus an inverse transform. From the same formula, or directly from (5.9), we also find that $u * v = v * u$, that is, that the convolution product of two n-tuples of complex numbers is commutative. Formula (5.13) can also be written:

$$F^{-1}(u * v) = nF^{-1}(u) \cdot F^{-1}(v),$$

that is, up to a factor n, *the Fourier transform of the convolution product is the componentwise product of the transforms of the factors.*

The convolution product of two n-tuples u and v arises as follows. Consider the two n-tuples as the n-tuples of the coefficients of two polynomials u and v in w of degree $n - 1$:

$$u = u_0 + u_1 w + \cdots + u_{n-1} w^{n-1},$$
$$v = v_0 + v_1 w + \cdots + v_{n-1} w^{n-1}.$$

Keeping in mind that $w^n = 1$, the product uv of the two polynomials is the polynomial in w having again degree $n - 1$ and, as coefficients, the c_ks from (5.9).

The product of two polynomials in an indeterminate x can also be obtained as a convolution. However, this cannot be accomplished directly: if g and h are two polynomials of degree $n - 1$, the convolution of the n-tuples of their coefficients is again an n-tuple, while the product gh is a polynomial of degree $2n - 2$, so its coefficients form a $(2n - 1)$-tuple. So we resort to a stratagem: consider g and h as polynomials of virtual degree $2n - 2$, by adding to them $n - 1$ monomials $0 \cdot x^n$, $0 \cdot x^{n+1}$, ..., $0 \cdot x^{2n-2}$:

$$u = (u_0, u_1, \ldots, u_{n-1}, u_n, u_{n+1}, \ldots, u_{2n-2}),$$
$$v = (v_0, v_1, \ldots, v_{n-1}, v_n, v_{n+1}, \ldots, v_{2n-2}),$$

with $u_i = v_i = 0$, for $i \geq n$. From (5.7), with $2n - 1$ for n, we have all terms $u_i v_j$ with $i + j \geq 2n - 1$ equal to zero, since at least one among i and j has to be $\geq n$ (else $i + j < 2n - 2$), so the corresponding u_i or v_j is zero. So the formula (5.7) becomes $c_k = \sum_{i+j=k} u_i v_j$, which is the formula giving the k-th coefficient of the polynomial product of g and h. In this way, we obtain the product of two polynomials from the convolution of their coefficients.

Example. Let us compute the product of two integer numbers by using convolution. Let the two numbers be 21 and 16. Write $21 = 1 \cdot 10^0 + 2 \cdot 10$ and $16 = 6 \cdot 10^0 + 1 \cdot 10$. Since the product has three digits, we have the two triples $u = (1, 2, 0)$ and $v = (6, 1, 0)$. The relevant Fourier matrix is F_3. We have:

$$F_3^{-1}(u) = \frac{1}{3} \begin{bmatrix} 3 \\ 1 + 2w^2 \\ 1 + 2w \end{bmatrix}, \quad F_3^{-1}(v) = \frac{1}{3} \begin{bmatrix} 7 \\ 6 + w^2 \\ 6 + w \end{bmatrix}.$$

The componentwise product yields $z = \frac{1}{9}[21, 4 + 11w^2, 4 + 11w]$, so

$$3 \cdot F_3(z) = \frac{1}{3}\begin{bmatrix} 18 \\ 39 \\ 6 \end{bmatrix} = \begin{bmatrix} 6 \\ 13 \\ 2 \end{bmatrix},$$

that is, $6 \cdot 10^0 + 13 \cdot 10^1 + 2 \cdot 10^2 = 6 \cdot 10^0 + 3 \cdot 10 + 3 \cdot 10^2 = 336$.

Let us see now how the coefficients y_k of the convolution product $z * x$ given by (5.11) can be found by performing a suitable linear transformation on z. If v_s are the Fourier coefficients of $x = Fv$ we have:

$$nv_s = x_0 + x_1\overline{w}^s + x_2\overline{w}^{2s} \cdots + \cdots + x_{n-1}\overline{w}^{(n-1)s},$$

or, equivalently,

$$nv_s = x_0 + x_{n-1}w^s + x_{n-2}w^{2s} + \cdots + x_1w^{(n-1)s}.$$

Set $x_0 = a_0$, $x_{n-i} = a_i$; we have:

$$nv_s = a_0 + a_1w^s + a_2w^{2s} + \cdots + a_{n-1}w^{(n-1)s}.$$

If:

$$f(x) = a_0 + a_1x + \cdots + a_{n-1}x^{n-1}, \tag{5.14}$$

then:

$$y_k = n\sum_{s=0}^{n-1} u_s v_s w^{ks} = \sum_{s=0}^{n-1} nv_s \cdot u_s w^{ks} = \sum_{s=0}^{n-1} f(w^s)u_s \cdot w^{ks}.$$

The terms $f(w^s)u_s$ are the components of the vector Az obtained by transforming z by the matrix:

$$A = \begin{pmatrix} a_0 & a_1 & a_2 & \cdots & a_{n-1} \\ a_{n-1} & a_0 & a_1 & \cdots & a_{n-2} \\ \vdots & \vdots & \vdots & \ddots & \vdots \\ a_1 & a_2 & a_3 & \cdots & a_0 \end{pmatrix}. \tag{5.15}$$

So the result contained in (5.11) can be expressed as follows:

Theorem 5.3. *If $u_0, u_1, \ldots, u_{n-1}$ are the Fourier coefficients of z, and if $z' = Az$, where A is matrix (5.15), then the Fourier coefficients of z' are:*

$$u_0f(1), \ u_1f(w), \ \ldots, \ u_sf(w^s), \ \ldots, \ u_{n-1}f(w^{n-1}),$$

where $f(x)$ is the polynomial (5.14).

5.3 Circulant matrices

A matrix like (5.15) is called *circulant*, since its rows are obtained by successive cyclic permutations of the first one. We write:

$$A = \mathrm{circ}(a_0, a_1, \ldots, a_{n-1}).$$

Note that a matrix $A = (a_{i,j})$ is cyclic if and only if $a_{i,j} = a_{i+1,j+1}$ (indices mod n). By Theorem 5.3, if A is such a matrix, then $u' = F^{-1}z' = F^{-1}Az = F^{-1}AF(u) = \{u_s f(w^s)\}$, that is:

$$F^{-1}AF \begin{bmatrix} u_0 \\ u_1 \\ \vdots \\ u_{n-1} \end{bmatrix} = \begin{pmatrix} f(1) & & & \\ & f(w) & & \\ & & \ddots & \\ & & & f(w^{n-1}) \end{pmatrix} \begin{bmatrix} u_0 \\ u_1 \\ \vdots \\ u_{n-1} \end{bmatrix},$$

and since this holds for every n-tuple u and every circulant matrix A, Theorem 5.3 can be stated as follows:

Theorem 5.4. *The Fourier matrix $F = F_n$ diagonalises all $n \times n$ circulant matrices:*

$$F^{-1}AF = \begin{pmatrix} f(1) & & & \\ & f(w) & & \\ & & \ddots & \\ & & & f(w^{n-1}) \end{pmatrix}, \tag{5.16}$$

where $A = \mathrm{circ}(a_0, a_1, \ldots, a_{n-1})$ and $f(x)$ is the polynomial (5.14).

We have seen that the columns $v_0, v_1, \ldots, v_{n-1}$ of the matrix F form a basis of the vector space of dimension n on complex numbers, C^n; so F is the change of basis matrix from the basis $\{(1, 0, \ldots, 0), (0, 1, 0, \ldots, 0), \ldots, (0, 0, \ldots, 0, 1)\}$ of the space C^n to the basis $\{v_i\}$. Hence Theorem 5.4 says that if ϕ is the linear transformation that in the standard basis is represented by the matrix A then, in the basis $\{v_i\}$, ϕ is represented by the diagonal matrix (5.16). In other words, $\phi(v_i) = f(w^i)v_i$, that is, the eigenvectors of any circulant matrix A are the columns v_j of the Fourier matrix (i.e. they do not depend on A), and the eigenvalues corresponding to $v_0, v_1, \ldots, v_{n-1}$ are $f(1), f(w), \ldots, f(w^{n-1})$, respectively, where $f(x)$ is the polynomial (5.14).

Remarks. 1. The numbers $f(w^i)$ may not be all distinct.

2. The matrix M of the v_is of (5.8) is circulant.

A converse of Theorem 5.4 holds:

Theorem 5.5. *Let $D = \mathrm{diag}\{\lambda_0, \lambda_1, \ldots, \lambda_{n-1}\}$ be a diagonal matrix (with the λ_i not necessarily all distinct). Then the matrix $A = FDF^{-1}$ is circulant.*

Proof. There exists a unique polynomial $f(x)$ of degree at most $n - 1$ whose value at w^i is λ_i. By Theorem 5.4, the circulant matrix $A = \mathrm{circ}(a_0, a_1, \ldots, a_{n-1})$, where the a_is are the coefficients of $f(x)$, is such that:

$$F^{-1}AF = \mathrm{diag}\{f(1), f(w), \ldots, f(w^{n-1})\}.$$

But this diagonal matrix is exactly the matrix D, and the claim follows. $\diamond$

From (5.16) we have that $\det A = \prod_{i=0}^{n-1} f(w^i)$ and, since the numbers w^i are the roots of $x^n - 1$, we have:

Theorem 5.6. *The determinant of a circulant matrix is the resultant*

$$\det(A) = R(x^n - 1, f(x)) = a_{n-1}^n \prod_{i=0}^{n-1} (\alpha_i{}^n - 1),$$

where $f(x)$ is the polynomial (5.14) and α_i its roots.

Let now $A = \mathrm{circ}(a_0, a_1, \ldots, a_{n-1})$ and $B = \mathrm{circ}(b_0, b_1, \ldots, b_{n-1})$ be two circulant matrices. The first row of the product AB is the n-tuple $(c_0, c_1, \ldots, c_{n-1})$, where:

$$c_0 = a_0 b_0 + a_1 b_{n-1} + a_2 b_{n-2} + \cdots + a_{n-1} b_1,$$
$$c_1 = a_0 b_1 + a_1 b_0 + a_2 b_{n-1} + \cdots + a_{n-1} b_2,$$
$$\vdots$$
$$c_{n-1} = a_0 b_{n-1} + a_1 b_{n-2} + a_2 b_{n-3} + \cdots + a_{n-1} b_0.$$

Keeping in mind (5.8), we recognise here the convolution $a * b$. The entries in the second row of AB are:

$$c_0' = a_{n-1} b_0 + a_0 b_{n-1} + a_1 b_{n-2} + \cdots + a_{n-2} b_1,$$
$$c_1' = a_{n-1} b_1 + a_0 b_0 + a_1 b_{n-1} + \cdots + a_{n-2} b_2,$$
$$\vdots$$
$$c_{n-1}' = a_{n-1} b_{n-1} + a_0 b_{n-2} + a_1 b_{n-3} + \cdots + a_{n-2} b_0,$$

so $c_0' = c_{n-1}$, $c_1' = c_0$, $c_2' = c_1$, $\ldots$, $c_{n-1}' = c_{n-2}$. Hence the first two rows of AB are:

$$AB = \begin{pmatrix} c_0 & c_1 & c_2 & \cdots & c_{n-1} \\ c_{n-1} & c_0 & c_1 & \cdots & c_{n-2} \\ \vdots & \vdots & \vdots & \vdots & \vdots \end{pmatrix}.$$

Analogously, $c_0'' = c_{n-2}$, $c_1'' = c_{n-1}$, $\ldots$, $c_{n-1}'' = c_{n-3}$, and so on. So we conclude that AB is the circulant matrix $\mathrm{circ}(c_0, c_1, \ldots, c_{n-1})$.

Thus:

Theorem 5.7. *The product of two complex circulant matrices is again circulant. Moreover, this product is commutative since so is convolution.*

A special role among circulant matrices is played by the matrix:

$$\sigma = \begin{pmatrix} 0 & 1 & 0 & \cdots & 0 \\ 0 & 0 & 1 & \cdots & 0 \\ \vdots & \vdots & \vdots & \ddots & \vdots \\ 0 & 0 & 0 & 0 & 1 \\ 1 & 0 & 0 & \cdots & 0 \end{pmatrix}.$$

If $u = (u_0, u_1, \ldots, u_{n-1})$ is a vector, we have $\sigma u = (u_1, u_2, \ldots, u_{n-1}, u_0)$; so σ determines on the components the cyclic permutation (which we denote by the same symbol σ used for the matrix):

$$\sigma = \begin{pmatrix} 0 & 1 & 2 & \ldots & n-2 & n-1 \\ 1 & 2 & 3 & \ldots & n-1 & 0 \end{pmatrix},$$

which we also write as $\sigma = (0, 1, 2, \ldots, n-1)$. Note that $\sigma^{-1} = \sigma^{n-1}$, both for the matrix and the permutation, and that for the matrix $\sigma^{-1} = \sigma^t$. Moreover, for all matrix $A = (a_{i,j})$ we have $\sigma^{-1} A \sigma = (a_{i+1,j+1})$ (indices taken mod n), since $A\sigma$ can be obtained from A by permuting the columns according to the permutation σ (the first column takes the place of the second one, the second one that of the third one and so on): so the entry $a_{i,j}$ goes to the position $(i, j+1)$. Left multiplication by σ^{-1} permutes the rows according to the same principle. So $a_{i,j}$ becomes $a_{i+1,j+1}$ in $\sigma^{-1} A\sigma$.

Since σ is circulant, its powers are too (Theorem 5.7). Now, if A is an arbitrary circulant matrix, we have $A\sigma = \sigma A$, since any two circulant matrices commute. Conversely, let A be a matrix that commutes with σ; then $\sigma^{-1} A\sigma = A$, so, by the above, $a_{i,j} = a_{i+1,j+1}$. In other words, A *is circulant if and only if it commutes with* σ. This fact can be seen to be a particular case of the following one. Since the powers of σ are obtained from σ by moving upwards the diagonal of ones, it is easy to see that if $A = \text{circ}(a_0, a_1, \ldots, a_{n-1})$, then A is a polynomial in σ:

$$A = a_0 + a_1\sigma + a_2\sigma^2 + \cdots + a_{n-1}\sigma^{n-1},$$

and conversely, if A has this form, since every summand is circulant, so is A. Hence: *A is circulant if and only if it is a polynomial in* σ.

For the matrix σ, the polynomial (5.14) is $f(x) = x$; so from (5.16) we get that *the eigenvalues of* σ *are the n-th roots of unity* $1, w, \ldots, w^{n-1}$.

5.4 The Fast Fourier transform

Evaluating a polynomial at a point requires, using Horner's method (Ch. 1, § 1.3), n multiplications; so, if we want the values of the polynomial at n points, n^2 multiplications are needed with this method. Interpolating at n points also requires a number of multiplications of the order of n^2, as does convolution, as (5.8) shows. If the n points are the n-th roots of unity $1, w, \ldots, w^{n-1}$, interpolation reduces, as seen in Theorem 5.2, to evaluating a polynomial at n points $1, \overline{w}, \ldots, \overline{w}^{n-1}$, and so does, by (5.13), convolution, since Fourier transform amounts to evaluating a polynomial at the points $\overline{w}^k$. Now, if n is a power of 2, $n = 2^l$, this computation can be arranged in such a way that the number of operations[3] is reduced from n^2 to $n \log n$. This is the reason why the procedure we shall now discuss is called *fast Fourier transform* (FFT). Note that the n-tuple $1, \overline{w}, \ldots, \overline{w}^{n-1}$ coincides with the n-tuple $1, w^{n-1}, \ldots, w$, so evaluating at the points $\overline{w}^k$ is equivalent to evaluating at w^k, up to order of the values.

So we are looking for the values of a polynomial of degree $n - 1$:

$$f(x) = a_0 + a_1 x + a_2 x^2 + \cdots + a_{n-1} x^{n-1}$$

at the points $1, w, w^2, \ldots, w^{n-1}$, with $n = 2^l$. We have to remark that, since n is even, the square of an n-th root of unity is a $\frac{n}{2}$-th root, since

$$\left((w^k)^2 \right)^{\frac{n}{2}} = w^{kn} = 1.$$

To take advantage of this fact, write the polynomial distinguishing even and odd powers:

$$f(x) = (a_0 + a_2 x^2 + \cdots + a_{n-2} x^{2\frac{n-2}{2}}) + x(a_1 + a_3 x^2 + \cdots + a_{n-1} x^{2\frac{n-2}{2}})$$
$$= f_1(x^2) + x f_2(x^2) = f_1(y) + x f_2(y),$$

so reducing the evaluation of $f(x)$ at the n-th roots to the evaluation of two polynomials $f_1(y)$ and $f_2(y)$, of degree $\frac{n}{2} - 1$, at the $\frac{n}{2}$-th roots of unity, plus n multiplications (of x by $f_2(y)$). Actually, there are just $n/2$ of the latter multiplications, since $w^{\frac{n}{2}} = 1$, so the n roots are $1, w, w^2, \ldots, w^{\frac{n}{2}-1}, -1, -w, -w^2, \ldots, -w^{\frac{n}{2}-1}$, and when the first $n/2$ products are known, the remaining ones are the same taken with the minus sign.

Since n is a power of 2, this procedure can be repeated for f_1 e f_2, obtaining:

$$f(x) = f_{11}(z) + y f_{12}(z) + x(f_{21}(z) + y f_{22}(z)),$$

and so on. In this way we arrive at 2^{l-1} polynomials of prime degree, and next to $n = 2^l$ polynomials of 0 degree, that is, the constants a_i coefficients of

---

[3] By "operations" we shall mean multiplications.

the initial polynomial $f(x)$, to be evaluated at the first roots of unity, that is, at 1. Being constants, the value is always a_i, so there are no more operations to be performed.

Example. Let $n = 8$.

$$
\begin{aligned}
f(x) &= a_0 + a_1 x + a_2 x^2 + a_3 x^3 + \cdots + a_7 x^7 \\
&= (a_0 + a_2 x^2 + a_4 x^4 + a_6 x^6) + x(a_1 + a_3 x^2 + a_5 x^4 + a_7 x^6) \\
&= a_0 + a_4 x^4 + x^2(a_2 + a_6 x^4) + x(a_1 + a_5 x^4 + x^2(a_3 + a_7 x^4)) \\
&= a_0 + x^4(a_4) + x^2(a_2 + x^4(a_6)) + x(a_1 + x^4(a_5) + x^2(a_3 + x^4(a_7))).
\end{aligned}
$$

So we have two polynomials of degree $3 = \frac{8}{2} - 1$ in x^2; having set $x^2 = y$ we have:

$$
f_1(y) = a_0 + a_2 y + a_4 y^2 + a_6 y^3, \quad f_2(y) = a_1 + a_3 y + a_5 y^2 + a_7 y^3.
$$

Each polynomial splits into two polynomials of degree $1 = \frac{8}{4} - 1$ in y^2 (so in x^4); setting $z = y^2$, we have:

$$
f_{11}(z) = a_0 + a_4 z, \ f_{12}(z) = a_2 + a_6 z, \ f_{21}(z) = a_1 + a_5 z, \ f_{22} = a_3 + a_7 z,
$$

and each of these polynomials splits into two polynomials of degree $0 = \frac{8}{8} - 1$; all in all, there are eight of them, and these are the eight coefficients of the polynomial. So we may write:

$$
f(x) = a_0 + a_4 z + y(a_2 + a_6 z) + x(a_1 + a_5 z + y(a_3 + a_7 z)).
$$

For every value of $x = 1, w, \ldots, w^{n-1}$, $z = x^4$ is a square root of unity, so $z = \pm 1$ (and both values are taken: for $x = w^k$ we have $z = \pm 1$ according to k being even or odd). We have to perform 4 multiplications by each of -1 and 1, so 8 multiplications in all (but see below). Similarly, y is a fourth root, so $y = 1, -1, i, -i$. For each of these values of y we have a value of $z = y^2$ so a single product $y(a_2 + a_6 z)$; similarly for $y(a_3 + a_7 z)$, so in all we have 8 products. We are left with the multiplication by x, to be performed 8 times, one for each root of unity. In total, $8 \cdot 3 = 24$ multiplications, or 8 multiplications for each of the three levels in which the computation has been divided.

In general, in order to go from a polynomial of degree $n - 1$, that is with n coefficients, to n polynomials of degree 0, that is, with a single coefficient, by dividing by 2 each time, the number of required steps is $\log n$. As seen in the example, it is necessary to perform n operations for each step, so the cost amounts to $n \log n$ operations. We shall come back to this problem in the next section.

It is to be noted that the cost just considered applies to the general scheme, rather than to actual computations. In the example, the number of operations is far smaller than 24. For one thing, the multiplications by the square roots of unity 1 and -1 are not to be performed; it suffices to take the coefficient

a to be multiplied or its opposite $-a$. In the computation that precedes these multiplications contributed 14 operations: indeed, 1 and -1 are square, fourth and eighth roots of unity, so they occur twice for z (so 8 times altogether), twice for y (4 times) and twice for x. So we go down from 24 to 10 operations. Multiplications by $-i$ appear 3 times (2 as a fourth root and 1 as an eighth root), and since the corresponding z is the same as for i, it is sufficient to take the opposite of the product by i, saving three operations. Analogously, we may avoid the multiplications by $w^5 = -w$ and $w^7 = -w^3$. Finally, we are left with the 5 operations:

$$w(a_1 - a_5), w^3(a_1 - a_5), i(a_2 - a_6), i(a_3 - a_7), i(a_1 + a_5).$$

Let us sum up the computation for $n = 8$:

$$f(1) = a_0 + a_4 + a_2 + a_6 + a_1 + a_5 + a_3 + a_7,$$
$$f(w) = a_0 - a_4 + i \cdot (a_2 - a_6) + w \cdot ((a_1 - a_5) + i \cdot (a_3 - a_7)),$$
$$f(w^2) = a_0 + a_4 - 1 \cdot (a_2 + a_6) + i \cdot ((a_1 + a_5) - 1 \cdot (a_3 + a_7)),$$
$$f(w^3) = a_0 - a_4 - i \cdot (a_2 - a_6) + w^3 \cdot ((a_1 - a_5) - i \cdot (a_3 - a_7)),$$

and starting from $f(w^4)$ we find again values we already know, keeping in mind that $w^4 = -1$, $w^5 = w^4 w = -w$, $w^6 = w^4 w^2 = -i$, and $w^7 = -w^3$:

$$f(w^4) = a_0 + a_4 + a_2 + a_6 - 1 \cdot (a_1 + a_5 + a_3 + a_7),$$
$$f(w^5) = a_0 - a_4 + i \cdot (a_2 - a_6) - w \cdot ((a_1 - a_5) + i \cdot (a_3 - a_7)),$$
$$f(w^6) = a_0 + a_4 - 1 \cdot (a_2 + a_6) - i \cdot ((a_1 + a_5) - 1 \cdot (a_3 + a_7)),$$
$$f(w^7) = a_0 - a_4 - i \cdot (a_2 - a_6) - w^3 \cdot ((a_1 - a_5) - i \cdot (a_3 - a_7)).$$

The order in which the coefficients of the polynomial appear at the end of the procedure is the one obtained from the initial ordering acting on the indices of $a_0, a_1, \ldots, a_{n-1}$ with the so-called *bit reversing permutation*. It can be obtained as follows. Write the numbers $0, 1, \ldots, n-1$ in base 2, i.e. using the binary digits (*bits*) 0 and 1, and then read each number in reverse order. For instance, for $n = 8$, the numbers:

$$0, 1, 2, 3, 4, 5, 6, 7,$$

are written:

$$000, 001, 010, 011, 100, 101, 110, 111.$$

By reversing the bits we get:

$$000, 100, 010, 110, 001, 101, 011, 111,$$

that is, the sequence:

$$0, 4, 2, 6, 1, 5, 3, 7,$$

as seen in the example.

The bit reversing permutation, which we shall denote by P_n for n numbers, can be obtained recursively as follows (*Buneman's algorithm*): having obtained P_n, the permutation P_{2n} can be written by doubling the numbers in the order given by P_n (so obtaining the first n numbers, the even ones) and adding 1 to each of them (obtaining the remaining numbers, the odd ones). So, starting from $n = 2$,

$$\underbrace{0\ 1}_{}$$
$$\downarrow \times 2$$
$$\underbrace{0\ 2}\ \overset{+1}{\rightarrow}\ 1\ 3$$
$$\downarrow \times 2$$
$$\underbrace{0\ 4\ 2\ 6}\ \overset{+1}{\rightarrow}\ 1\ 5\ 3\ 7$$
$$\downarrow \times 2$$
$$\underbrace{0\ 8\ 4\ 12\ 2\ 10\ 6\ 14}\ \overset{+1}{\rightarrow}\ 1\ 9\ 5\ 13\ 3\ 11\ 7\ 15$$

and so on. That the algorithm works can be seen as follows. Let m an integer, $m \in \{0, 1, \ldots, n-1\}$, and let:

$$m = a_1 a_2 \cdots a_r$$

be the binary expression of m ($a_i = 0, 1$). Hence:

$$P(m) = a_r \cdots a_2 a_1,$$

and doubling:

$$2P(m) = a_r \cdots a_2 a_1 0,$$

which is the inverse of

$$m = 0 a_1 a_2 \cdots a_r, \tag{5.17}$$

considering m as an element of $\{0, 1, \ldots, 2n-1\}$. Note that, since they start with 0, the numbers (5.17) describe the first n numbers in $\{0, 1, \ldots, 2n-1\}$. Moreover:

$$2P(m) + 1 = a_r \cdots a_2 a_1 1,$$

which is the inverse of:

$$m = 1 a_1 a_2 \cdots a_r,$$

describes the last n numbers in $\{0, 1, \ldots, 2n-1\}$.

The elements a_i that appear in the output of the algorithm are the coefficients of the powers x^i (with the same i). To reduce to the form containing only the powers $\frac{n}{2}, \frac{n}{4}, \ldots, 1$ of x, write the polynomial in the order given by the algorithm and factor out a power x^k in the sum of the monomials $a_h x^h$ with $h > k$ that follow $a_k x^k$. For instance, with $n = 16$, we have:

$$a_0 + a_8 x^8 + a_4 x^4 + \cdots + a_{15} x^{15},$$

and factoring out as described we have:

$$a_0 + a_8 x^8 + x^4(a_4 + a_{12}x^8) + x^2(a_2 + a_{10}x^8 + a_6 x^4 + a_{14}x^{12})$$
$$+ x(\dots),$$

where the dots represent the sum of the previous row with the indices incremented by one. Repeat next the procedure, if needed, within the parentheses (now the monomials are not $a_h x^h$, but $a_h x^{h-k}$); so for the term that multiplies x^2 we have:

$$a_2 + a_{10}x^8 + x^4(a_6 + a_{14}x^8).$$

5.5 $n \log n$ complexity

To measure the efficiency of an algorithm the *worst case* is often studied, that is, the case in which the running time is greatest ("time" refers here to the number of operations to be executed). By this approach, one tries to find out the relationship between the algorithm and its input, as a function $g(n)$ of the number of inputs, ignoring possible constants arising from the properties of particular inputs, from running conditions (characteristics of the computer, for instance), or from the actual program, which may be more or less clever. The order of growth of the function $g(n)$ is the *complexity* of the algorithm.

The scheme to evaluate a polynomial at the roots of unity seen above belongs to the class of algorithms that divide a problem into smaller problems, independent of each other, repeating this subdivision for each of these problems, up to a set of problems whose solution is trivial[4]. To find the solution of the original problem, the partial solutions are next combined with a cost which is, at each step, linear (at each step the data are considered just once).

So, in our case, C_n, the complexity for an input of magnitude $n = 2^l$ (the n coefficients of the polynomial to be evaluated at the n roots of unity), equals twice the complexity for the computation for one polynomial with $n/2$ coefficients (the polynomials f_1 and f_2), plus n multiplications (for the n roots of unity:

$$C_n = 2C_{\frac{n}{2}} + n,$$

with $C_1 = 0$ (evaluating constants at 1). Iterating the procedure with f_1 and f_2 we have the same equality with $n/2$ in lieu of n: $C_{\frac{n}{2}} = 2C_{\frac{n}{4}} + \frac{n}{2}$ which, substituted in the previous one, gives $C_n = 4C_{\frac{n}{4}} + 2n$, and in general $C_n = 2^k C_{\frac{n}{2^k}} + kn$. Now, $n = 2^l$, and with $k = l$ we have:

$$C_n = 2^k C_1 + l \cdot 2^l = l \cdot 2^l,$$

---

[4] The technique of subdividing a problem in independent smaller problems is known as "divide et impera" ("divide and conquer").

since $C_1 = 0$. From this follows:

$$C_n = 2^l l = n \log n.$$

In a problem where at each step it is necessary to consider just one half of the inputs, as is the case for the evaluation of a polynomial at the roots of unity, the computation just seen yields the result $\frac{1}{2} n \log n$.

As we have seen, the number of operations in a particular case can be smaller than $n \log n$. Saying that a given problem can be solved by performing $n \log n$ operations, or that it is a problem of $n \log n$ *complexity*, we are stating that it belongs to the class of problems for whose solution a scheme such as the above applies, a scheme involving in fact $n \log n$ operations. Fourier transform is one such problem.

References

[AHU] Chapter 7, [BCLR] Chapter4 , [Se] Chapter 41, [St] § 4.2 and 5.5, [Da], and [VL] (the last mentioned contains an ample bibliography). See [Se] and [BCLR] for more about algorithmic complexity.

5.6 Appendix

The discrete Fourier transform is related in a natural way to the theory of (finite) Abelian groups and their algebras (to be defined below). The *representation theory* of these groups is basically a possible interpretation of this transform, as we shall see.

5.6.1 Group algebras

Let $G = \{g_0 = 1, g_1, \ldots, g_{n-1}\}$ be a (not necessarily Abelian) finite group of order n, $\mathcal{C}$ the complex field, and let $\mathcal{C}[G]$ be the set of all functions

$$u : G \longrightarrow \mathcal{C}.$$

$\mathcal{C}[G]$ gets a vector space structure by defining:

$$(u + v)(g) = u(g) + v(g),$$
$$(\alpha u)(g) = \alpha(u(g)), \ \alpha \in \mathcal{C}.$$

The zero vector of this space is the function u equal to zero at each $g \in G$. A basis of $\mathcal{C}[G]$ is given by the n functions u_g defined as follows:

$$u_g(h) = \begin{cases} 1, \text{ if } h = g; \\ 0, \text{ otherwise.} \end{cases}$$

Indeed, every $u \in \mathcal{C}[G]$ can be written as:

$$u = \sum_{g \in G} u(g) u_g, \tag{5.18}$$

and if such a linear combination equals the zero vector, all the coefficients are zero, since $0 = u(h) = \sum_{g \in G} u(g) u_g(h) = u(g)$. So we have a vector space having as its dimension the order of the group G.

We can define on $\mathcal{C}[G]$ a product as follows. If $u, v \in \mathcal{C}[G]$, then let uv be the element of $\mathcal{C}[G]$ that takes at $g \in G$ the value so defined: for all $h \in G$ consider $h^{-1}g$ and set:

$$uv(g) = \sum_{h \in G} u(h) v(h^{-1}g). \tag{5.19}$$

Clearly, if $\alpha \in \mathcal{C}$ then $\alpha(uv) = (\alpha u)v = u(\alpha v)$. Moreover, as is easily verified, the associative and distributive properties hold, and an identity element exists: the function whose value is 1 at the group identity and zero elsewhere. So this is an algebra, the *group algebra* of the group G.

Consider now the map $G \longrightarrow \mathcal{C}[G]$ defined by $g \to u_g$. Note that

$$u_g u_h(t) = \sum_{s \in G} u_g(s) u_h(s^{-1}t),$$

and, since if $s \neq g$ or $s^{-1}t \neq h$ the summands are zero, only the term for $s = g$ and $s^{-1}t = h$ are left, that is, $g^{-1}t = h$. In other words,

$$u_g u_h(t) = \begin{cases} 1, \text{ if } t = gh; \\ 0, \text{ otherwise} \end{cases}$$

so, by definition, $u_g u_h = u_{gh}$. Therefore, the elements u_g combine as the elements g of G, so we may embed G in $\mathcal{C}[G]$ identifying g with u_g. Hence, $\mathcal{C}[G]$ contains a copy of G, and (5.18) can be written:

$$u = \sum_{g \in G} u(g) g. \tag{5.20}$$

In this way $\mathcal{C}[G]$ become the set of formal linear combinations of elements of G, and the product (5.19) extends to these linear combinations the product of G (*extension by linearity*). More precisely, with u as in (5.20) and $v = \sum_{h \in G} v(h) h$, we have $uv = \sum_{g,h \in G} u(g) v(h) gh$. The uv-coefficient of the element s of G, when uv is written in the form (5.20), is that obtained for $gh = s$, that is, for $h = g^{-1}s$; so it is $\sum_{g \in G} u(g) v(g^{-1}s)$, or the value defined above. Thus we have:

$$uv = \sum_{s} \left(\sum_{g} u(g) v(g^{-1}s) \right) s. \tag{5.21}$$

Remark. Note that from $gh = s$ the relation $g = sh^{-1}$ also follows, so:

$$\sum_{g \in G} u(g)v(g^{-1}s) = \sum_{h \in G} u(sh^{-1})v(h).$$

The product (5.21) is the *convolution product*. We find again the product seen in § 5.3 by considering the case when the group G is cyclic, $G = \{1, g, g^2, \ldots, g^{n-1}\}$. Indeed, if we write u_k for $u(g^k)$, and $u = \sum_{i=0}^{n-1} u_i g^i$, $v = \sum_{j=0}^{n-1} v_j g^j$, the coefficient of g^k in the product uv according to (5.21) is $\sum_{i=0}^{n-1} u_i v_{k-i}$, as in (5.11).

5.6.2 Cyclic groups

The Fourier transform interpreted in the group algebra $\mathcal{C}[G]$ of a cyclic group G is simply a change of basis with respect to the basis $G = \{1, g, g^2, \ldots, g^{n-1}\}$. The aim of this change is for the convolution product of two elements, written in the new basis, to be carried out componentwise.

More precisely, we have, as a vector space,

$$\mathcal{C}[G] = \mathcal{C} \cdot 1 \oplus \mathcal{C} \cdot g \oplus \cdots \oplus \mathcal{C} \cdot g^{n-1},$$

and this space get an algebra structure by the convolution product (5.21). By writing $u = \alpha_0 \cdot 1 + \alpha_1 \cdot g + \cdots + \alpha_{n-1} \cdot g^{n-1}$ and $v = \beta_0 \cdot 1 + \beta_1 \cdot g + \cdots + \beta_{n-1} \cdot g^{n-1}$, in the product uv the terms $\alpha_i \beta_j g^i g^j$ $(i \neq j)$ appear, that is products belonging to distinct components. Let us look for a new basis $e_0, e_1, \ldots, e_{n-1}$:

$$\mathcal{C}[G] = \mathcal{C}e_0 \oplus \mathcal{C}e_1 \oplus \cdots \oplus \mathcal{C}e_{n-1}, \tag{5.22}$$

such that $e_i e_j = 0$ for $i \neq j$, so by writing u and v as above the terms $\alpha_i \beta_j e_i e_j$, $i \neq j$, are zero. Thus, the convolution product will be simply given by the products of the same-name components.

So, transform the basis $\{g^i\}$ by the matrix $F^{-1} = \frac{1}{n}\overline{F}$, and let $\{e_i\}$ be the basis so obtained. Hence, e_i is the element of $\mathcal{C}[G]$ that, written in the form (5.20), has as its g^j-coefficient the complex number $\overline{w}^{ij}/n$:

$$e_i = \sum_{j=0}^{n-1} \frac{\overline{w}^{ij}}{n} g^j.$$

Note that

$$e_i g = w^i e_i, \tag{5.23}$$

and that from this identity $e_i e_j = 0$ follows for $i \neq j$. Indeed, since $e_i e_j g = e_i g e_j$, we have $e_i w^j e_j = w^i e_i e_j$, and hence $(w^i - w^j)e_i e_j = 0$, and for $i \neq j$ we have $w^i \neq w^j$, so:

$$e_i e_j = 0, \quad i \neq j. \tag{5.24}$$

Moreover,

$$\sum_{i=0}^{n-1} e_i = \frac{1}{n} \sum_{j=0}^{n-1} \left(\sum_{i=0}^{n-1} \overline{w}^{ij} \right) g^j,$$

and since the inner sum equals 0 for $j \neq 0$ and n for $j = 0$,

$$e_0 + e_1 + \cdots + e_{n-1} = 1. \tag{5.25}$$

Multiplying the last identity by e_i and keeping in mind (5.24), we have:

$$e_i^2 = e_i, \tag{5.26}$$

for all $i = 0, 1, \ldots, n - 1$. So the e_is are orthogonal idempotents summing to 1. Note that by summing (5.23) over i, we have $(\sum e_i)g = g = \sum w^i e_i$, or:

$$g = 1 \cdot e_0 + w \cdot e_1 + \cdots + w^{n-1} e_{n-1},$$

and analogously:

$$g^k = 1 \cdot e_0 + w^k \cdot e_1 + \cdots + w^{k(n-1)} e_{n-1},$$

which is also clear from the fact that the change of basis matrix from $\{e_i\}$ to the basis $\{g^k\}$ is the inverse matrix of F^{-1}, that is, F. In general, for

$$u = u_0 + u_1 g + \cdots + u_{n-1} g^{n-1}$$

we have from (5.23):

$$u = \sum_{i=0}^{n-1} e_i u = \sum_{i=0}^{n-1} (\sum_{k=0}^{n-1} u_k w^{ik}) e_i = \sum_{i=0}^{n-1} z_i e_i,$$

(cf. (5.7)), and if analogously $v = \sum_{i=0}^{n-1} x_i e_i$ the product (5.21) of u and v is

$$uv = (z_0 x_0) e_0 + (z_1 x_1) e_1 + \cdots + (z_{n-1} x_{n-1}) e_{n-1}.$$

So, (5.22) is a decomposition of $\mathcal{C}[G]$ in a direct sum of rings (in the sense discussed in Chapter 1). Each summand is isomorphic to the complex field $\mathcal{C}$ since the correspondence $\mathcal{C} \longrightarrow \mathcal{C}e_i$, while clearly preserving the sum, also preserves the product since if $\alpha \to \alpha e_i$ and $\beta \to \beta e_i$, then $\alpha\beta \to \alpha\beta e_i^2 = \alpha\beta e_i$, by (5.26). To sum up, *the group algebra of a cyclic group is a direct product of algebras isomorphic to the complex field*. This is another way of conveying the meaning of Fourier transform.

The circulant matrices belong to the picture depicted in this section. Indeed, they are the matrices of the linear transformations of the space $\mathcal{C}[G]$ obtained by multiplying the basis $\{1, g, \ldots, g^{n-1}\}$ by a given element u of $\mathcal{C}[G]$. If $u = g$, we obtain the transformation which, in this basis, is represented by the matrix σ seen in § 5.3; multiplying by the powers of g, we get the powers of σ. If $u = \sum_{i=0}^{n-1} a_i g^i$, we get the matrix (5.15).

5.6.3 The character group

Let us see now how to extend what we have seen in the last two sections to the case of an arbitrary Abelian group G, that is, not necessarily a cyclic one. Let us see first a necessary condition. If for the group algebra $\mathcal{C}[G]$ there is a decomposition such as (5.22), with e_i orthogonal idempotents, then if $g = \sum_i \alpha_i(g)e_i$ and $h = \sum_i \alpha_i(h)e_i$, we have $gh = \sum_i \alpha_i(g)\alpha_i(h)e_i$. But the element gh is written as $gh = \sum_i \alpha_i(gh)e_i$, so, since $\{e_i\}$ is a basis, gh is written in a unique way and

$$\alpha_i(gh) = \alpha_i(g)\alpha_i(h), \qquad (5.27)$$

for all i and $g, h \in G$. In other words, the correspondence associating with an element g of G its i-th coefficient in the expression $g = \sum_i \alpha_i(g)e_i$ is a homomorphism $G \longrightarrow \mathcal{C}^*$, the multiplicative group of the complex field $\mathcal{C}$. (Note that we cannot have $\alpha_i(g) = 0$ since, as $1 = 1 \cdot e_0 + 1 \cdot e_1 + \cdots + 1 \cdot e_{n-1}$, we have $\alpha_i(1) = 1$ for all i, so if $\alpha_i(g) = 0$ for some i we would have $1 = \alpha_i(1) = \alpha_i(gg^{-1}) = \alpha_i(g)\alpha_i(g^{-1}) = 0 \cdot \alpha_i(g^{-1}) = 0$.) In the case in which G is cyclic, we have $\alpha_i(g^k) = w^{ik}$, and (5.27) is verified.

A homomorphism $\chi : G \longrightarrow \mathcal{C}^*$ is called a *character* of the Abelian group G, and the set of the characters of G is itself a group, with respect to the operation

$$(\chi\chi')(g) = \chi(g)\chi'(g).$$

Clearly, this operation is associative. The identity element is the character χ_1 whose value is 1 at every element of G (*principal* or *trivial* character), while the inverse of a character χ is the character whose value at g is the value χ takes at g^{-1}: $\chi^{-1}(g) = \chi(g^{-1})$. This group is the *character group* of G; it is denoted as $\widehat{G}$.

The following important result relies on the fundamental theorem of finite Abelian groups (that is, that any finite Abelian group is a direct product of cyclic groups).

Theorem 5.8 (Duality theorem). *For a finite Abelian group G, the groups G and $\widehat{G}$ are isomorphic.*

Proof. If G is cyclic, $G = \langle g \rangle$, a homomorphism is completely determined when the image of g is known. If g has order n, then $\chi(g)^n = 1$ since $\chi(g)^n = \chi(g^n) = \chi(1) = 1$. So $\chi(g)$ is an n-th root of unity and, there being n of them, we have n possible homomorphisms, that is, n characters. If χ_k is the character determined by the root w^k, where w is a primitive root, that is, $\chi_k : g \to w^k$, the correspondence $G \longrightarrow \widehat{G}$ given by

$$g^k \to \chi_k$$

is the isomorphism we are looking for. In the general case, G is a direct product of cyclic groups, $G = G_1 \times G_2 \times \cdots \times G_t$, with $|G_i| = n_i$ and $G_i = \langle g_i \rangle$.

Let g be a fixed element of G: it admits a unique expression $g = g_1^{k_1} g_2^{k_2} \cdots g_t^{k_t}$. Let $h = g_1^{h_1} g_2^{h_2} \cdots g_t^{h_t}$ be a generic element of G and w_i a primitive n_i-th root of unity. Define:

$$\chi_g : h \to w_1^{k_1 h_1} w_2^{k_2 h_2} \cdots w_t^{k_t h_t}.$$

It is clear that $\chi_{gg'} = \chi_g \chi_{g'}$. If $g \neq g'$, then $k_i \neq k_i'$ for at least one i, so $\chi_g(g_i) = w_i^{k_i} \neq w_i^{k_i'} = \chi_{g'}(g_i)$. As g spans G, all the characters χ_g are distinct, and the correspondence $G \longrightarrow \widehat{G}$ given by:

$$g \to \chi_g$$

is the promised isomorphism. Note that $\chi_g(h) = \chi_h(g)$. ◇

Remark. Similarly, $\widehat{G} \simeq \widehat{\widehat{G}}$, so $G \simeq \widehat{G}$ e $G \simeq \widehat{\widehat{G}}$. But while the first isomorphism is not "natural", in the sense that to define it it is necessary to choose a "basis" of G (a g_i for each factor G_i) and to write the elements in this basis, the second one can be explicitly defined on the elements:

$$g \to \widehat{\chi_g} : \chi \to \chi(g),$$

that is, by associating to $g \in G$ the character of $\widehat{G}$ that maps a character χ of G to the value it takes at g.

Let us discuss some properties of the characters.

Lemma 5.1. *For every character χ, the following holds:*

$$\sum_g \chi(g) = \begin{cases} n, & \text{if } \chi = \chi_1; \\ 0, & \text{otherwise.} \end{cases}$$

Proof. If $\chi = \chi_1$, the result is straightforward. Otherwise, there is an element g' such that $\chi(g') \neq 1$. So, since gg', for g in G, spans the whole set of elements of G, we have:

$$\sum_g \chi(g) = \sum_g \chi(gg') = \sum_g \chi(g)\chi(g') = \chi(g') \sum_g \chi(g);$$

hence $(1 - \chi(g')) \sum_g \chi(g) = 0$, and since $\chi(g') \neq 1$, we have $\sum_g \chi(g) = 0$. ◇

By duality, we get:

Lemma 5.2. *For every element g, the following holds:*

$$\sum_{\chi \in \widehat{G}} \chi(g) = \begin{cases} n, & \text{if } g = 1; \\ 0, & \text{otherwise.} \end{cases}$$

Proof. In the previous lemma, substitute g for χ, 1 for χ_1, and χ' for g'. ◇

Corollary 5.1. *For all elements g, h, the following holds:*

$$\sum_{\chi \in \widehat{G}} \chi(g)\chi(h^{-1}) = \begin{cases} n, & \text{if } g = h; \\ 0, & \text{otherwise.} \end{cases}$$

Lemma 5.3 (orthogonality relations). *We have:*

$$\sum_{g} \chi_i(g)\chi_j(g^{-1}) = \begin{cases} n, & \text{if } i = j; \\ 0, & \text{otherwise.} \end{cases}$$

Proof. If $i = j$, then $\chi_i(g)\chi_i(g^{-1}) = \chi_i(1) = 1$, and the result follows. If $i \neq j$, there is h such that $\chi_i(h) \neq \chi_j(h)$; so,

$$\sum_{g} \chi_i(g)\chi_j(g^{-1}) = \sum_{g} \chi_i(gh)\chi_j(h^{-1}g^{-1})$$

$$= \left(\sum_{g} \chi_i(g)\chi_j(g^{-1})\right)(\chi_i(h)\chi_j(h^{-1})),$$

and hence:

$$(1 - \chi_i(h)\chi_j(h^{-1}))\sum_{g} \chi_i(g)\chi_j(g^{-1}) = 0.$$

If $\chi_i(h)\chi_j(h^{-1}) = 1$, then we have $\chi_i(h) = \chi_j(h)$, against the choice of h. Hence the second factor is the one equal to zero, as required.   ◇

Remark. If G is cyclic and generated by g, we have $\chi_i(g) = w^i$, where w is an n-th primitive root of unity. The sum in the previous lemma can then be written as $\sum_{k=1}^{n-1} \chi_i(g^k)\chi_j(g^{-k})$, and the lemma reduces to Theorem 5.1.

Writing the elements of G as $g_1, g_2, \ldots, g_n$ and, correspondingly, the characters as $\chi_1, \chi_2, \ldots, \chi_n$, we may consider the $n \times n$ matrix whose (i, j) entry is $\chi_i(g_j)$. This matrix is the *character table*; we shall denote it by T. As remarked at the end of the proof of Theorem 5.1, we have $\chi_i(g_j) = \chi_j(g_i)$, so T is symmetric. In the case of a cyclic group, the character table is the Fourier matrix.

Define now in $\mathcal{C}[G]$ an inner product as follows:

$$(u, v) = \frac{1}{n}\sum_{g} u(g)v(g^{-1}).$$

The $\chi(g)$s being roots of unity implies that $\chi(g^{-1}) = \chi(g)^{-1} = \overline{\chi(g)}$. Hence, Lemma 5.3 states that among the elements of $\mathcal{C}[G]$ the characters form an *orthonormal* system:

$$(\chi_i, \chi_j) = 0, \; i \neq j; \quad (\chi_i, \chi_i) = 1.$$

In particular, they are linearly independent and, as there are n of them, *the characters form a basis of $C[G]$*, so an orthonormal basis. Hence the product defined above is non-degenerate, and the character table T turns out to be a non-singular matrix (its determinant equals ± 1). Moreover, as in the case of the Fourier matrix, $T^{-1} = \frac{1}{n}\overline{T}$.

Example. Let $G = \{1, a, b, c\}$ be the Klein (four-)group: every element has period 2, and the product of two elements different from 1 equals the third one. In this case, since $g^2 = 1$ for every element of G, the $\chi_i(g)$s are second roots of unity, so they are either -1 or 1. There are four homomorphisms $G \longrightarrow C$: the kernel of one of them, χ_1, is the whole G, while the other three have kernels $\{1, a\}$, $\{1, b\}$, and $\{1, c\}$; we shall denote them as χ_2, χ_3, and χ_4, respectively. So we have the table:

$$
\begin{array}{c}
\\
\chi_1 \\
\chi_2 \\
\chi_3 \\
\chi_4
\end{array}
\begin{array}{cccc}
1 & a & b & c \\
\left(\begin{array}{cccc}
1 & 1 & 1 & 1 \\
1 & 1 & -1 & -1 \\
1 & -1 & 1 & -1 \\
1 & -1 & -1 & 1
\end{array}\right)
\end{array}.
$$

5.6.4 The algebra of an Abelian group

As we have seen at the beginning of the previous section, a necessary condition for an element $g \in G$ to admit an expression of the form $g = \sum_i \alpha_i(g)e_i$, as is the case for cyclic groups, is that the coefficients $\alpha_i(g)$ are values taken by characters. Let us see now how to find such an expression, first of all by defining the elements e_i, $i = 1, 2, \ldots, n$.[5] Having chosen an ordering $\chi_1, \chi_2, \ldots, \chi_n$ of the characters, define, as in the cyclic case:

$$
e_i = \sum_g \frac{\chi_i(g^{-1})}{n} g \quad \left(= \sum_g \frac{\overline{\chi_i(g)}}{n} g \right).
$$

The e_is so defined are orthogonal idempotents summing to 1. The proof is analogous to that for cyclic groups. Indeed, we have:

$$
\sum_{i=1}^{n} e_i = \frac{1}{n} \sum_g \left(\sum_i \chi_i(g^{-1}) \right) g,
$$

and by Lemma 5.2 the inner sum equals 0 for $g \neq 1$, and equals n if $g = 1$. From this follows

$$
e_1 + e_2 + \cdots + e_n = 1.
$$

---

[5] Following the prevailing custom we denote here the principal character by χ_1 and not by χ_0; consequently, in this section we label the e_is starting with e_1, rather than with e_0 as we have previously done.

Moreover, for all $h \in G$,

$$e_i h = \chi_i(h) e_i, \tag{5.28}$$

since:

$$e_i h = \sum_g \frac{\chi_i(g^{-1})}{n} gh = \sum_s \frac{\chi_i(hs^{-1})}{n} s,$$

where we have set $gh = s$, and

$$\chi_i(h) e_i = \chi_i(h) \sum_g \frac{\chi_i(g^{-1})}{n} g = \sum_g \frac{\chi_i(hg^{-1})}{n} g.$$

Hence it follows:

$$e_i e_j = 0, \ i \neq j \, ;$$

indeed, $e_i e_j h = \chi_j(h) e_i e_j$, so $e_i h e_j = \chi_i(h) e_i e_j$. The left-hand sides being equal, the right-hand ones are too, so $(\chi_j(h) - \chi_i(h)) e_i e_j = 0$. But for at least one element h of G we have $\chi_j(h) \neq \chi_i(h)$ for $i \neq j$, so $e_i e_j = 0$. From this relation and from $\sum_i e_i = 1$ the idempotence property $e_i^2 = e_i$ follows.

Thus the elements e_i form a basis for the group algebra, so we have here too a decomposition (5.22). The change from the basis $\{g_i\}$ to the basis $\{e_i\}$ is provided by the linear transformation having as its matrix the inverse of the character table. A group element g can be written, summing (5.28) over i, as

$$g = \chi_1(g) e_1 + \chi_2(g) e_2 + \cdots + \chi_n(g) e_n, \tag{5.29}$$

and a generic element $u = \sum_g u(g) g$ of $\mathcal{C}[G]$ as $\sum_i z_i e_i$, where the complex number z_i equals $\sum_g u(g) \chi_i(g)$.

Remark. Since there are $n!$ ways of ordering the n characters of G, there are $n!$ ways of decomposing $\mathcal{C}[G]$ as seen.

The correspondence $\sum_i z_i e_i \to (z_1, z_2, \ldots, z_n)$ yields an isomorphism between the group algebra $\mathcal{C}[G]$ and the direct sum of n copies of the complex field $\mathcal{C}$:

$$\mathcal{C}[G] \simeq \mathcal{C} \oplus \mathcal{C} \oplus \cdots \oplus \mathcal{C}, \tag{5.30}$$

and there are $n!$ ways of establishing such an isomorphism. The decomposition (5.30) only depends on n, that is, only on the group order and not on its structure. In other words, *the group algebras of the Abelian groups of order n are all isomorphic*. In particular, given an Abelian group G, in its group algebra a copy of every other Abelian group of the same order can be found. More precisely, let G' be an Abelian group of order $n = |G|$, $\mathcal{C}[G']$ its group algebra, $\chi'_1, \chi'_2, \ldots, \chi'_n$ the characters G' ordered in an arbitrary way, h an element of G' and

$$(\chi'_1(h), \chi'_2(h), \ldots, \chi'_n(h))$$

the n-tuple corresponding to the expression (5.29) of h in $\mathcal{C}[G']$, that is, the n-tuple corresponding to the element h in the isomorphism (5.30) for the algebra $\mathcal{C}[G']$. Now associate with h the element u of $\mathcal{C}[G]$, having as its i-th component in the e_i-basis the complex number $\chi'_i(h)$:

$$h \to u = \sum_i \chi'_i(h)e_i.$$

It is clear that if k is another element of G', and $v \in \mathcal{C}[G]$ is its image, then the image of the product hk is the product uv of the images. Moreover, the map above is injective since in the character table there are no equal columns, as it is a non-singular matrix. Thus the image of G' in $\mathcal{C}[G]$ is a group isomorphic to G'.

If we want the expression of u in the form $\sum_g u(g)g$, which amounts to knowing the image under u of every $g \in G$, we may proceed as follows. We know that $u = \sum_i (\sum_g u(g)\chi_i(g))e_i$; so we have to determine the $u(g)$s in such a way that

$$\sum_g u(g)\chi_i(g) = \chi'_i(h),$$

for $i = 1, 2, \ldots, n$. Hence we have a system of n linear equations in the n unknowns $u(g)$, $g \in G$, and the matrix of this system is the character table. The $u(g)$s can then be determined using Cramer's rule.

Example. Let G be the Klein group and G' the cyclic group $C_4 = \{1, x, x^2, x^3\}$. The character table of the former has been seen in the previous example; that of the latter is the Fourier matrix:

$$
\begin{array}{c}
\\
\chi_1 \\
\chi_2 \\
\chi_3 \\
\chi_4
\end{array}
\begin{array}{cccc}
1 & x & x^2 & x^3 \\
\left(\begin{array}{cccc}
1 & 1 & 1 & 1 \\
1 & i & -1 & -i \\
1 & -1 & 1 & -1 \\
1 & -i & -1 & i
\end{array}\right)
\end{array}.
$$

Let now x be the element h as above; the 4-tuple corresponding to it is

$$x \to (1, i, -1, -i),$$

and if $x \to u \in \mathcal{C}[G]$, then the $u(g)$s, $g \in G$, are determined by the system:

$$
\begin{cases}
u(1)\chi_1(1) + u(a)\chi_1(a) + u(b)\chi_1(b) + u(c)\chi_1(c) = 1, \\
u(1)\chi_2(1) + u(a)\chi_2(a) + u(b)\chi_2(b) + u(c)\chi_2(c) = i, \\
u(1)\chi_3(1) + u(a)\chi_3(a) + u(b)\chi_3(b) + u(c)\chi_3(c) = -1, \\
u(1)\chi_4(1) + u(a)\chi_4(a) + u(b)\chi_4(b) + u(c)\chi_4(c) = -i,
\end{cases}
$$

or,

$$
\begin{cases}
u(1) + u(a) + u(b) + u(c) = 1, \\
u(1) + u(a) - u(b) - u(c) = i, \\
u(1) - u(a) + u(b) - u(c) = -1, \\
u(1) - u(a) - u(b) + u(c) = -i.
\end{cases}
$$

It is straightforward to see that this system admits the solution:

$$u(1) = u(b) = 0, \; u(a) = \frac{1+i}{2}, \; u(c) = \frac{1-i}{2},$$

that determines u in $\mathcal{C}[G]$. So we have:

$$u = 0 \cdot 1 + \frac{1+i}{2}a + 0 \cdot b + \frac{1-i}{2}c = \frac{1+i}{2}a + \frac{1-i}{2}c.$$

Analogously, for x^2 we find, if $x^2 \to v$,

$$v(1) = v(a) = v(c) = 0, \; v(b) = 1,$$

that is,

$$v = 0 \cdot 1 + 0 \cdot a + 1 \cdot b + 0 \cdot c = 1 \cdot b,$$

and for x^3, if $x^3 \to z$,

$$z(1) = z(b) = 0, \; z(a) = \frac{1-i}{2}, \; z(c) = \frac{1+i}{2},$$

or,

$$z = 0 \cdot 1 + \frac{1-i}{2}a + 0 \cdot b + \frac{1+i}{2}c = \frac{1-i}{2}a + \frac{1+i}{2}c.$$

Check, for instance, that the image of the product $x \cdot x^2$, i.e. the image of x^3, that is z, is actually equal to the product of the images of x and x^2, that is, uv; this element, computed using (5.21), yields:

$$s = 1: u(1)v(1) + u(a)v(a) + u(b)v(b) + u(c)v(c) = 0,$$
$$s = a: u(1)v(a) + u(a)v(1) + u(b)v(c) + u(c)v(b) = \tfrac{1-i}{2},$$
$$s = b: u(1)v(b) + u(a)v(c) + u(b)v(1) + u(c)v(a) = 0,$$
$$s = c: u(1)v(c) + u(a)v(b) + u(b)v(a) + u(c)v(1) = \tfrac{1+i}{2}.$$

So we have:

$$uv = \frac{1-i}{2}a + \frac{1+i}{2}c,$$

which is precisely z.

Remark. The fact that the group algebra G contains a copy of every Abelian group of order equal to $|G|$ is analogous to the fact that, once a group of order n has been embedded in the symmetric group S^n (Cayley's theorem), copies of every group of order n can be found in S^n.

We have seen that the characters χ_i of an Abelian group G make up a basis for the vector space $\mathcal{C}[G]$. So every $u \in \mathcal{C}[G]$ can be written in a unique way, as:

$$u = \sum_i c_i \chi_i. \tag{5.31}$$

The coefficient c_j of χ_j can be computed by taking the inner product of u by χ_j:

$$(u, \chi_j) = \frac{1}{n} \sum_g u(g)\overline{\chi_j(g)} = \frac{1}{n} \sum_g \sum_i c_i(\chi_i, \chi_j)$$

$$= \sum_i c_i \left(\frac{1}{n} \sum_g \chi_i(g)\overline{\chi_j(g)} \right)$$

$$= c_j,$$

where the last equality follows from the orthogonality relations. Thus the coefficient c_j of χ_j in u is:

$$c_j = \frac{1}{n} \sum_g u(g)\overline{\chi_j(g)}. \tag{5.32}$$

If $u = u_g$, that is, the function taking value equal to 1 at g and zero elsewhere, which we have identified with $g \in G$, (5.32) becomes:

$$c_j = \frac{\overline{\chi_j(g)}}{n},$$

and (5.31):

$$g = \sum_i \frac{\overline{\chi_i(g)}}{n}\chi_i.$$

Note that the coefficients c_i are exactly those necessary to write the idempotents of the basis $\{e_i\}$ in the basis $\{g_i\}$, as seen at the start of this chapter:

$$e_i = \sum_{g \in G} c_i g,$$

the change of base matrix for $\{g_i\} \to \{e_i\}$ being $\frac{1}{n}\overline{T} = T^{-1}$. By (5.31),

$$g = \sum_{i=1}^{n} c_i \chi_i.$$

Here the sum is over i, not, as above, over g. So rows and columns are exchanged in the c_i-matrix, and the resulting matrix turns out to be the transposed matrix $(T^{-1})^t$. But since T is symmetric, T^{-1} is too, so the matrix is the same one.

References

[A]   Ascoli G.: Lezioni di Algebra. Editrice Tirrena, Torino (1965).
[AHU]   Aho A.V., Hopkroft J.E., Ulmann J.D.: The Design and Analysis of Computer Algorithms. Addison-Wesley, Boston (1974).
[C]   Childs L.: A Concrete Introduction to Higher Algebra. Springer-Verlag, Berlin Heidelberg New York (1983).
[CA]   Buchberger B.: Computer Algebra. Symbolic and Algebraic Computation, 2nd edition. Springer-Verlag, Berlin Heidelberg New York (1983).
[CMP]   Cerlienco L., Mignotte M., Piras F.: Suites récurrentes linéaires. L'Enseignement Mathématique **33**: 67–108 (1987).
[Da]   Davis P.J.: Circulant Matrices. Wiley, New York (1979).
[DST]   Davenport J., Siret Y., Tournier E.: Calcul formel. Masson, Paris (1987); English translation: Computer Algebra: Systems and Algorithms for Algebraic Computation. Academic Press, New York (1988).
[Kn]   Knuth D.E.: The Art of Computer Programming, II: Seminumerical Algorithms, 2nd edition. Addison-Wesley, Boston (1981).
[Ku]   Kurosh A.G.: Higher algebra. MIR, Moscow (1972).
[Li]   Lipson J.D.: Chinese remainder and interpolation algorithms. Proceedings of the second ACM symposium on Symbolic and algebraic manipulation, SYMSAC '71, pp. 372–391.
[LLL]   Lenstra A.K., Lenstra H.W., Lovász L.: Factoring polynomials with rational coefficients. Math. Ann., **261**: 515–534 (1982).
[LN]   Lidl R., Niederreiter H.: Finite Fields. Encyclopedia of Mathematics and its Applications. Addison-Wesley, Boston (1983).
[McE]   McEliece R.J.: Finite Fields for Computer Scientists and Engineers. Kluwer (1987).
[Mi]   Mignotte M.: Mathématiques pour le calcul formel. PUF, Paris (1989); English translation: Mathematics for Computer Algebra. Springer-Verlag, Berlin Heidelberg New York (1992).
[Se]   Sedgewick R.: Algorithms, 2nd edition. Addison-Wesley, Boston (1988).
[St]   Strang G.: Introduction to Applied Mathematics. Wellesley-Cambridge University Press (1986).
[VL]   Van Loan C.: Computational Frameworks for the Fast Fourier Transform. SIAM, Philadelphia (1992).

Index

Collana Unitext – La Matematica per il 3+2

Series Editors:
A. Quarteroni (Editor-in-Chief)
L. Ambrosio
P. Biscari
C. Ciliberto
G. van der Geer
G. Rinaldi
W.J. Runggaldier

Editor at Springer:
F. Bonadei
francesca.bonadei@springer.com

As of 2004, the books published in the series have been given a volume number.
Titles in grey indicate editions out of print.
As of 2011, the series also publishes books in English.

A. Bernasconi, B. Codenotti
Introduzione alla complessità computazionale
1998, X+260 pp, ISBN 88-470-0020-3

A. Bernasconi, B. Codenotti, G. Resta
Metodi matematici in complessità computazionale
1999, X+364 pp, ISBN 88-470-0060-2

E. Salinelli, F. Tomarelli
Modelli dinamici discreti
2002, XII+354 pp, ISBN 88-470-0187-0

S. Bosch
Algebra
2003, VIII+380 pp, ISBN 88-470-0221-4

S. Graffi, M. Degli Esposti
Fisica matematica discreta
2003, X+248 pp, ISBN 88-470-0212-5

S. Margarita, E. Salinelli
MultiMath - Matematica Multimediale per l'Università
2004, XX+270 pp, ISBN 88-470-0228-1

A. Quarteroni, R. Sacco, F.Saleri
Matematica numerica (2a Ed.)
2000, XIV+448 pp, ISBN 88-470-0077-7
2002, 2004 ristampa riveduta e corretta
(1a edizione 1998, ISBN 88-470-0010-6)

13. A. Quarteroni, F. Saleri
    Introduzione al Calcolo Scientifico (2a Ed.)
    2004, X+262 pp, ISBN 88-470-0256-7
    (1a edizione 2002, ISBN 88-470-0149-8)

14. S. Salsa
    Equazioni a derivate parziali - Metodi, modelli e applicazioni
    2004, XII+426 pp, ISBN 88-470-0259-1

15. G. Riccardi
    Calcolo differenziale ed integrale
    2004, XII+314 pp, ISBN 88-470-0285-0

16. M. Impedovo
    Matematica generale con il calcolatore
    2005, X+526 pp, ISBN 88-470-0258-3

17. L. Formaggia, F. Saleri, A. Veneziani
    Applicazioni ed esercizi di modellistica numerica
    per problemi differenziali
    2005, VIII+396 pp, ISBN 88-470-0257-5

18. S. Salsa, G. Verzini
    Equazioni a derivate parziali – Complementi ed esercizi
    2005, VIII+406 pp, ISBN 88-470-0260-5
    2007, ristampa con modifiche

19. C. Canuto, A. Tabacco
    Analisi Matematica I (2a Ed.)
    2005, XII+448 pp, ISBN 88-470-0337-7
    (1a edizione, 2003, XII+376 pp, ISBN 88-470-0220-6)

20. F. Biagini, M. Campanino
    Elementi di Probabilità e Statistica
    2006, XII+236 pp, ISBN 88-470-0330-X

21. S. Leonesi, C. Toffalori
    Numeri e Crittografia
    2006, VIII+178 pp, ISBN 88-470-0331-8

22. A. Quarteroni, F. Saleri
    Introduzione al Calcolo Scientifico (3a Ed.)
    2006, X+306 pp, ISBN 88-470-0480-2

23. S. Leonesi, C. Toffalori
    Un invito all'Algebra
    2006, XVII+432 pp, ISBN 88-470-0313-X

24. W.M. Baldoni, C. Ciliberto, G.M. Piacentini Cattaneo
    Aritmetica, Crittografia e Codici
    2006, XVI+518 pp, ISBN 88-470-0455-1

25. A. Quarteroni
    Modellistica numerica per problemi differenziali (3a Ed.)
    2006, XIV+452 pp, ISBN 88-470-0493-4
    (1a edizione 2000, ISBN 88-470-0108-0)
    (2a edizione 2003, ISBN 88-470-0203-6)

26. M. Abate, F. Tovena
    Curve e superfici
    2006, XIV+394 pp, ISBN 88-470-0535-3

27. L. Giuzzi
    Codici correttori
    2006, XVI+402 pp, ISBN 88-470-0539-6

28. L. Robbiano
    Algebra lineare
    2007, XVI+210 pp, ISBN 88-470-0446-2

29. E. Rosazza Gianin, C. Sgarra
    Esercizi di finanza matematica
    2007, X+184 pp, ISBN 978-88-470-0610-2

30. A. Machì
    Gruppi – Una introduzione a idee e metodi della Teoria dei Gruppi
    2007, XII+350 pp, ISBN 978-88-470-0622-5
    2010, ristampa con modifiche

31 Y. Biollay, A. Chaabouni, J. Stubbe
    Matematica si parte!
    A cura di A. Quarteroni
    2007, XII+196 pp, ISBN 978-88-470-0675-1

32. M. Manetti
    Topologia
    2008, XII+298 pp, ISBN 978-88-470-0756-7

33. A. Pascucci
    Calcolo stocastico per la finanza
    2008, XVI+518 pp, ISBN 978-88-470-0600-3

34. A. Quarteroni, R. Sacco, F. Saleri
    Matematica numerica (3a Ed.)
    2008, XVI+510 pp, ISBN 978-88-470-0782-6

35. P. Cannarsa, T. D'Aprile
    Introduzione alla teoria della misura e all'analisi funzionale
    2008, XII+268 pp, ISBN 978-88-470-0701-7

36. A. Quarteroni, F. Saleri
    Calcolo scientifico (4a Ed.)
    2008, XIV+358 pp, ISBN 978-88-470-0837-3

37. C. Canuto, A. Tabacco
    Analisi Matematica I (3a Ed.)
    2008, XIV+452 pp, ISBN 978-88-470-0871-3

38. S. Gabelli
    Teoria delle Equazioni e Teoria di Galois
    2008, XVI+410 pp, ISBN 978-88-470-0618-8

39. A. Quarteroni
    Modellistica numerica per problemi differenziali (4a Ed.)
    2008, XVI+560 pp, ISBN 978-88-470-0841-0

40. C. Canuto, A. Tabacco
    Analisi Matematica II
    2008, XVI+536 pp, ISBN 978-88-470-0873-1
    2010, ristampa con modifiche

41. E. Salinelli, F. Tomarelli
    Modelli Dinamici Discreti (2a Ed.)
    2009, XIV+382 pp, ISBN 978-88-470-1075-8

42. S. Salsa, F.M.G. Vegni, A. Zaretti, P. Zunino
    Invito alle equazioni a derivate parziali
    2009, XIV+440 pp, ISBN 978-88-470-1179-3

43. S. Dulli, S. Furini, E. Peron
    Data mining
    2009, XIV+178 pp, ISBN 978-88-470-1162-5

44. A. Pascucci, W.J. Runggaldier
    Finanza Matematica
    2009, X+264 pp, ISBN 978-88-470-1441-1

45. S. Salsa
    Equazioni a derivate parziali – Metodi, modelli e applicazioni (2a Ed.)
    2010, XVI+614 pp, ISBN 978-88-470-1645-3

46. C. D'Angelo, A. Quarteroni
    Matematica Numerica – Esercizi, Laboratori e Progetti
    2010, VIII+374 pp, ISBN 978-88-470-1639-2
    2012, ristampa con modifiche

47. V. Moretti
    Teoria Spettrale e Meccanica Quantistica – Operatori in spazi di Hilbert
    2010, XVI+704 pp, ISBN 978-88-470-1610-1

48. C. Parenti, A. Parmeggiani
    Algebra lineare ed equazioni differenziali ordinarie
    2010, VIII+208 pp, ISBN 978-88-470-1787-0

49. B. Korte, J. Vygen
    Ottimizzazione Combinatoria. Teoria e Algoritmi
    2010, XVI+662 pp, ISBN 978-88-470-1522-7

50. D. Mundici
    Logica: Metodo Breve
    2011, XII+126 pp, ISBN 978-88-470-1883-9

51. E. Fortuna, R. Frigerio, R. Pardini
    Geometria proiettiva. Problemi risolti e richiami di teoria
    2011, VIII+274 pp, ISBN 978-88-470-1746-7

52. C. Presilla
    Elementi di Analisi Complessa. Funzioni di una variabile
    2011, XII+324 pp, ISBN 978-88-470-1829-7

53. L. Grippo, M. Sciandrone
    Metodi di ottimizzazione non vincolata
    2011, XIV+614 pp, ISBN 978-88-470-1793-1

54. M. Abate, F. Tovena
    Geometria Differenziale
    2011, XIV+466 pp, ISBN 978-88-470-1919-5

55. M. Abate, F. Tovena
    Curves and Surfaces
    2011, XIV+390 pp, ISBN 978-88-470-1940-9

56. A. Ambrosetti
    Appunti sulle equazioni differenziali ordinarie
    2011, X+114 pp, ISBN 978-88-470-2393-2

57. L. Formaggia, F. Saleri, A. Veneziani
    Solving Numerical PDEs: Problems, Applications, Exercises
    2011, X+434 pp, ISBN 978-88-470-2411-3

58. A. Machì
    Groups. An Introduction to Ideas and Methods of the Theory of Groups
    2011, XIV+372 pp, ISBN 978-88-470-2420-5

59. A. Pascucci, W.J. Runggaldier
    Financial Mathematics. Theory and Problems for Multi-period Models
    2011, X+288 pp, ISBN 978-88-470-2537-0

60. D. Mundici
    Logic: a Brief Course
    2012, XII+124 pp, ISBN 978-88-470-2360-4

61. A. Machì
    Algebra for Symbolic Computation
    2012, VIII+174 pp, ISBN 978-88-470-2396-3

The online version of the books published in this series is available at SpringerLink.
For further information, please visit the following link:
http://www.springer.com/series/5418